California Natural History Guides: 43

GEOLOGIC HISTORY
OF MIDDLE CALIFORNIA

BY
ARTHUR D. HOWARD

UNIVERSITY OF CALIFORNIA PRESS
BERKELEY • LOS ANGELES • LONDON

California Natural History Guides
Arthur C. Smith, General Editor

Advisory Editorial Committee
Mary Lee Jefferds
A. Starker Leopold
Robert Ornduff
Robert C. Stebbins

University of California Press
Berkeley and Los Angeles, California
University of California Press, Ltd.
London, England

Library of Congress Catalog Card Number: 78-57299
Printed in the United States of America

ISBN 0–520–03874–6
2 3 4 5 6 7 8 9

CONTENTS

PREFACE

A forerunner of this book, entitled *Evolution of the Landscape of the San Francisco Bay Region,* was primarily concerned with the landscape of the nine counties adjoining San Francisco Bay during the last 10 million years or so. In this book, I have extended the area of coverage to the north and south, and considerably farther to the east, well into the foothills of the Sierra Nevada. In addition, I have projected the story much farther back in time. The early part of this story, up to 65 million years ago, is necessarily treated cursorily because details are hazier. From 65 million years ago to the present, however, evidence of the past becomes increasingly abundant and greater detail is possible. Because of new studies, we are also able to say more about special places of interest such as Clear Lake, Monterey Bay, Sutter Buttes, Point Reyes, and the Sierran foothills. A new series of sequential sketches has been prepared to illustrate the history of the area through time.

Our story is necessarily a compressed synthesis of the contributions of a great many individuals, including representatives of the U.S. Geological Survey, the California Division of Mines and Geology, other Federal and State agencies, local colleges and universities, and a number of private industries concerned with natural resources and the environment. The list is far too great for individual acknowledgements. A number of selected references are included for those wishing to pursue certain aspects of the story in more detail. I am indebted to the following colleagues for fruitful discussions: W. R. Dickinson, W. R. Evitt, J. C. Ingle, B. M. Page, and E. I. Rich. I am especially indebted to B. M. Page and E. I. Rich for critical reading of considerable parts of the text. I alone am responsible for errors in fact or oversimplification.

A. D. H.

INTRODUCTION

The landscape of the San Francisco Bay region and middle California represents the culmination of a remarkable series of natural events that began many millions of years ago. To these ancient events we owe not only the major topographic features of the landscape—the Coast Ranges, the Great Valley, and the Sierra Nevada—but also the innumerable smaller features that lend variety and charm to the region. Only by delving into the past can we satisfactorily account for the present beautiful setting of San Francisco Bay, for the picturesque ranges of the coastal belt, for the many lovely intermontane valleys and basins, for scenic Monterey Bay and Clear Lake, for the isolated Point Reyes Peninsula, for Sutter Buttes rising out of the plain north of Sacramento, for the spirelike topography of Pinnacles National Monument, and for a host of other fascinating landscape features.

In describing the evolution of the landscape, I will introduce events that may at first tax the credulity of the reader. I will probe the role of *plate tectonics,* [1] according to which the Earth's crust is divided into gigantic moving plates which jostle each other like huge ice floes and determine not only the configuration of continental borders, but the origin and distribution of great fractures like the San Andreas Fault and the localization of earthquakes and volcanoes. I will describe the upheaval and destruction of ancient mountain ranges, widespread invasions and withdrawals of the sea, the birth and extinction of volcanoes, and the changing climates of the past.

The concept of change is vital to our story. The landscape is not permanently set in stone; its appearance at any one time represents a fleeting episode in an endless struggle—a struggle in which powerful deforming forces, originating largely within the Earth, are arrayed against forces from without. The deforming forces crumple and break the outer shell of the Earth

---

1. *Tectonics* is a general term referring to deformation of the Earth's crust by fracturing, uplift, warping, and crumpling, singly or in combination—activities that commonly result in mountains.

and commonly heave it into mountains. Their activity is usually accompanied by volcanic activity and jarring earthquakes. We must not suppose, however, that the rise of mountains is cataclysmic; from the human standpoint it is painfully slow. It is true that in some regions, like the western United States, an entire mountain range may break loose from its surroundings and snap upward a few feet, but these movements rarely occur more than once a century. Thus, the rate of growth of even such mountains averages only a few inches a year. Other ranges rise without perceptible movement, by an infinitesimally slow buckling or crumpling of the ground. Precise surveys must be repeated periodically over long spans of time to reveal these faint changes in elevation. The cumulative effects of even these imperceptible movements are the lofty mountains of the Earth. This explains why we find entombed in the rocks of many high peaks the fossilized shells of creatures that once lived in the sea, thousands of feet below.

Against the powerful internal forces are arrayed others, much less spectacular, that operate more subtly and insidiously and pass largely unnoticed. These are the forces of weathering and erosion which slowly but relentlessly nibble away at the highlands created by the forces from within. When the internal forces dominate, mountains stand high and dot the surface of the Earth; when the internal forces are dormant, the processes of weathering and erosion dominate and carry the substance of the mountains into the sea. There have been times in the past when the internal forces lay dormant for such long periods of time that the mountains in some regions were completely obliterated and the landscape was reduced to a low plain.

The landscape, then, changes constantly at the whim of the forces that produce it. The present landscape may be likened to a single frame in a timeless motion picture—a motion picture run at an inconceivably slow rate.

It is our intent in this little book to rerun a small part of this film—a part amounting to about 2.5 percent of the great span of Earth history—only that part in which the unusual past of Middle California is involved. Before delving into this story, however, let us consider briefly the characteristics of the present landscape, for it is this landscape that we seek to explain.

THE MODERN LANDSCAPE

The overall landscape of middle California is relatively simple. It consists of three major topographic units: the Sierra Nevada, the Great Valley of California, and the Coast Ranges.

The Regional Setting

The Sierra Nevada, one of the world's great mountain ranges, forms a 400-mile-long (650-kilometers) barrier between the arid lands of Nevada and the Great Valley of California. Trapdoorlike in profile, the range presents a precipitous escarpment on the east — a rugged wall that must have dismayed the early pioneers seeking to reach the California gold fields. The crest of the escarpment is topped by a line of lofty peaks of which Mt. Whitney, the highest peak in the United States exclusive of Alaska, is perhaps the best known. The descent from the High Sierra westward to the Great Valley is gentle, encompassing almost the entire width of the range. This gentle slope is scarred by a number of deep canyons, of which Yosemite Valley is justly famous.

The Great Valley of California, comparable to the Sierra Nevada in geographic dimensions, is an almost level plain lying at or close to sea level. Actually, in the delta area east of Suisun Bay the surface is as much as 17 feet (5 meters) below sea level and is protected from the inroads of the sea and from river floods by an impressive network of levees and dikes. The landscape is strikingly reminiscent of many parts of Holland. The floor of the valley rises gently toward the flanking mountains, where it reaches altitudes of several hundred feet. The only landscape features of any prominence within the valley are Sutter Buttes, also known as Marysville Buttes, a domal uplift surmounted by a cluster of ancient volcanoes about 40 miles (65 kilometers) north of Sacramento (Photo 1).

The northern half of the Great Valley is drained by the Sacramento River, which rises in the Klamath Mountains far to the north and flows south into San Francisco Bay. The southern

1

half of the valley is drained by the San Joaquin River, which rises in the southern Sierra not far from the source of Yosemite's Merced River. The San Joaquin joins the Sacramento just east of San Francisco Bay.

The Coast Ranges are a great natural barrier between the Great Valley and the Pacific Ocean (Photo 2). The individual ranges vary considerably in size and shape. Some are poorly defined sprawling masses like the mountainous wilderness stretching north from the latitude of Clear Lake. Others are large, well-defined units separated by broad valleys.

The Coast Range belt varies in width from as little as 40 miles (65 kilometers) in the vicinity of San Francisco Bay to as much as 90 miles (145 kilometers) in the far north. The belt trends about 30 degrees west of north, roughly parallel to the coast, but many of the individual ranges trend obliquely across the belt and terminate abruptly at the coast on the west, and less abruptly at the Great Valley on the east. Interspersed among the ranges are numerous long, linear lowlands such as Petaluma Valley and the San Francisco Bay lowland, and a number of irregular basins such as Livermore Basin and the basin of Clear Lake. The Coast Ranges are not particularly high, averaging between 2000 and 4000 feet (600 and 1200 meters) with only a few peaks rising higher.

Until now we have been surveying the California landscape as though from the fringes of space. Let us descend now to lower altitudes and bring the middle California landscape into sharper focus.

The Local Setting

The larger landscape features of Middle California are pictorially shown in Figure 1. Notice that San Francisco Bay and its companions, San Pablo and Suisun bays, split the Coast Range belt in two, and that the mountain ranges to the north and south extend toward it like stubby fingers. Notice, too, that the ranges south of the bay are much larger than those immediately to the north.

To the south, the Coast Range belt consists of three major mountain blocks. These are, from east to west: the Diablo

Range, terminating northward in the Contra Costa Hills and the Diablo Hills; the Santa Cruz–Gabilan mountain tract; and the Santa Lucia Range along the coast. Between the Diablo Range and the Santa Cruz–Gabilan tract is a long lowland, flooded by San Francisco Bay in the north. For simplicity's sake we shall refer to this as the San Francisco Bay–Santa Clara lowland. The lowland between the Santa Cruz–Gabilan tract and the Santa Lucia Mountains we shall refer to as the Salinas lowland, after the river that drains it.

This simple picture of large mountain ranges and broad intervening lowlands does not apply north of San Francisco Bay. It is true that for 45 or 50 miles (70 or 80 kilometers) to the north there are linear mountains separated by lowlands, but they are small and irregular compared to those to the south. These small ranges, from east to west, are the Vaca, Howell, Mayacmas, Sonoma, and Marin. The intervening valleys, named in the same order, are the Berryessa-Suisun, Napa, Sonoma-Rincon, and Petaluma. The Berryessa-Suisun and Sonoma-Rincon are strings of basins rather than continuous valleys.

These northern ranges and lowlands lose their individuality within 45 or 50 miles of San Francisco Bay. Beyond that point they more or less merge into a broad, dissected upland, the Mendocino Plateau, the crest of which descends gradually seaward. Only the long depression occupied by Petaluma Creek and the Russian River penetrates northward into the plateau.

Four peaks dominate the Bay Region landscape. All are accessible by road and afford superb views of the surrounding territory. These are Mt. Hamilton in the Diablo Range, 4206 feet (1283 meters) high and site of the famous Lick Observatory; Mt. Diablo, 3849 feet (1174 meters) high, at the northeastern extremity of the Diablo Range (Photo 3); Mt. Tamalpais, 2606 feet (795 meters) high, in the Marin Mountains north of the Golden Gate; and Mt. St. Helena, 4336 feet (1332 meters) high, in the Mayacmas Mountains (Photo 4).

In contrast to the Coast Ranges, the Great Valley is essentially a featureless plain. Deep drilling, however, has revealed

3

Figure 1. Physiographic diagram of Middle California.

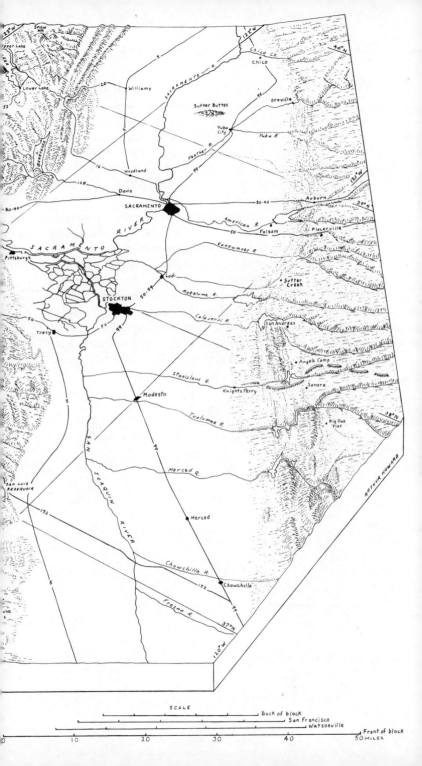

SCALE

Back of block
San Francisco
Watsonville
Front of block

0   10   20   30   40   50 MILES

a wealth of data on the sediments that underlie the valley. This data provides much of the basis for our reconstruction of the past history of the valley.

The Sierran foothills consist of low dissected plains adjacent to the Great Valley, and a steeper, increasingly dissected slope rising eastward toward the Sierran crest. In the south, the foothills are dominated by bold, discontinuous ridges roughly parallel to the margin of the Great Valley.

This, then, is the landscape for which we seek explanation. In any discussion of geologic history, a question that frequently arises is, "How do you know that such an event took place?" A very reasonable question, inasmuch as we are dealing with events that took place millions of years ago. Actually, the unraveling of ancient geologic history requires a special kind of training. The person so trained is a geologist. Geologists are in reality detectives, because from obscure clues scattered about the landscape they deduce events to which there have been no witnesses. In a sense, then, they emulate the fictional Sherlock Holmes. Let us see how this is done.

PROBING THE PAST

One of the fundamental truths of Nature is that the processes that are shaping the landscape today have acted in similar fashion throughout the long course of Earth history though not necessarily at the same rate. From the record preserved in the rocks, we know that streams eroded as relentlessly in the distant past as they do now; glaciers ground and gouged as do the Alpine and Antarctic glaciers of today; winds swept sand about, etched the faces of the rocks, and created fields of dunes; the waves of the sea battered ancient shorelines as they do the present California coast; volcanoes spewed lava and ashes as they do today; and the all-powerful internal forces heaved up mountains and caused earthquakes even as now.

Not only have the processes remained the same, but the products of their activity—the beds of clay, sand, and gravel, and the layers of lava and volcanic ash—are essentially the same, whether one or 100 million years old. It is this similarity of process and product throughout time that enables the geologist, the student of Earth history, to ferret out the story of the Earth. Rocks are the products of events. Could one read the rocks, the whole fascinating history of the Earth would be laid bare. To read the past, however, requires an understanding of the present; *the present is the key to the past.*

The layers of sediment that are spread over the lowlands of the Earth by streams, winds, and glacial ice, and those that are laid down on the sea floor by waves and currents, are like the pages of a history book. The sedimentary layer at the bottom of any sequence is, of course, the oldest and is covered in turn by successively younger ones. Thus, the layers of sediment in the walls of deep canyons, such as the Grand Canyon of the Colorado River or—on a smaller scale—Niles Canyon between the San Francisco Bay lowland and Livermore Basin, provide a record of the past going back many millions of years.

What have the rocks to teach us? First, they tell us of the ancient patterns of land and sea, of the geographies of ancient times. For example, sediments that are being deposited today beneath the sea entomb the remains of marine organisms, of clams and oysters, starfish and sea urchins, and myriad other marine creatures. In contrast, the sediments that are laid down on dry land, the continental sediments, entomb the remains of land plants and animals. Suppose, then, that our geological detective traces one of the rock layers that is widely exposed in the Coast Ranges. Suppose that in the eastern part of the Coast Range belt this formation contains the remains of land plants and animals, whereas in the western part the fossils are those of marine organisms. Obviously, then, the ancient shoreline lay somewhere between. If this procedure is repeated for each of the succeeding layers exposed in the Coast Ranges, a complete record of the shifting shorelines of the past emerges. It is even possible to determine, within limits, the temperature of the water of the ancient seas. This is done first by analogy — by observing the temperatures favored by the modern descendents of the marine creatures entombed in the ancient formations. The second method is by chemical analysis of the fossil shell material, because the exact composition depends on the temperature of the water in which the animal lived.

In practice, the deciphering of ancient geographies is a difficult, time-consuming procedure because of the vast size of the pages of our history book and because parts or whole pages have been removed by erosion. The deciphering of a single page, even in a relatively limited area, requires the investigative effort of scores of geologists and may take generations.

We might inquire next as to how the topography of the past is determined—that is, how are ancient mountains and lowlands located? Again, relying on the concept that the present is the key to the past, if we examine the sediments that are being eroded from present mountains and spread out over the adjacent lowlands, we come to some interesting conclusions. One is that as the mountain streams reach the lowlands and lose velocity, they are forced to deposit much of their sedimentary burden. Naturally, the largest fragments, the boulders and cobbles, are dropped first, and the sand, silt, and clay are carried farther. Thus, by examining the variations in particle size in an ancient sedimentary layer, the direction of the source of the sediment is easily determined. The volume of the sediment eroded from the mountains may provide an approximation of the size of the mountain mass; and the composition of the pebbles, cobbles, and boulders provides information on the rocks that composed them.

In many areas the geologist can even determine the source of the pressures that crumpled up ancient mountain ranges. This is rather difficult in the Coast Ranges because of the complicated geology. It is much simpler in the Appalachian Mountains of the eastern United States, where the layers of rock are crumpled into folds like wrinkles in a loose rug (Figure 2). The rock folds of the Appalachians become smaller and smaller to the west and eventually die out completely. Thus, the pressure must have come from the east. One can achieve the same result experimentally by moistening a piece of tissue paper on a glass plate and pushing one side toward the other. The wrinkles die out away from the source of the pressure. So it is frequently a simple matter to deduce the source of the pressures that crumpled up ancient mountain ranges. How these powerful forces were set in motion, however, is a more difficult question. The new concept of plate tectonics, however, is helping to resolve many hitherto perplexing problems. Before considering the specific role of plate tectonics in the evolution of coastal California, a few preparatory remarks on the structure and composition of the outer part of the Earth are advisable.

The Earth is composed of a number of shells, the outermost of which is the *lithosphere*. The lithosphere probably averages 45–60 miles (70–95 kilometers) thick; it is thinner under the

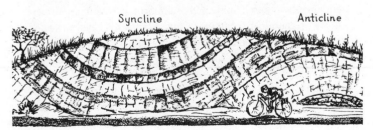

Figure 2. Rock folds: syncline and anticline.

oceans and thicker under the continents. The so-called *crust* of the Earth is the upper part of the lithosphere. Under the oceans, the crust consists of dark, solidified lava known as *basalt,* the same material that is brought up to the surface by the volcanoes of Hawaii. The continental crust, on the other hand, consists largely of the light-colored rock, *granite*, such as is exposed in the Sierra Nevada. Granite, too, was once molten but unlike lava, solidified miles below the surface of the Earth. We see it exposed now in deep canyons or in the cores of mountain ranges only because erosion has removed the rocks that once concealed it. Thus, the crust consists largely of the formerly molten *(igneous)* rocks, basalt and granite.

All rocks exposed to the weather and to erosion are broken down into loose particles of sediment and subject to solution by passing waters. The sedimentary particles are spread out in sheets by streams, wind, glaciers, and ocean currents and eventually solidify to rock. Much of the dissolved materials eventually end up in the sea, where they precipitate on the sea floor. Rocks formed by consolidation of sediments are known as *sedimentary rocks*. Thus, *conglomerates* are cemented gravel; *sandstone*, cemented sand; *shale*, solidified mud; and *limestone,* solidified lime mud. Because of deformation, the layers of sedimentary rock in many places are no longer in their original horizontal or near horizontal positions. Instead, they are inclined. The inclination is referred to as *dip*, and is measured in degrees from the horizontal. Where the edges of dipping layers appear at the surface, their direction across country is known as *strike* and is measured with a compass.

Both igneous and sedimentary rocks may be changed to *metamorphic rocks* if subjected to abnormal pressures, temperatures, or chemical solutions. Thus, shales may be converted to *slate* or, with increasing metamorphism, to a *phyllite* in which myriad tiny mica flakes impart a silky sheen to the rock, or to a *schist* (pronounced "shist") in which the mica flakes are large and easily visible. Sandstones are altered to the resistant rock *quartzite*; limestones are converted to *marble*; and sedimentary rocks of mixed composition may be changed to the metamorphic rock *gneiss* (pronounced "nice"). Many gneisses display alternating layers or stringers of light and dark minerals. Basalts are commonly metamorphosed to a dark schist, and granite, to a gneiss. Sedimentary rocks invaded by magma may be altered near the contact by temperature, fluids, and—to some extent—by pressures. As a result, they become hard and dense.

The three major classes of rock—igneous, sedimentary, and metamorphic—comprise the crust of the Earth. Basalt underlies the ocean floors but also occurs on the lands. Granite and metamorphic rocks, formed at high temperatures, are largely confined to the continents. Sedimentary rocks form a blanket over parts of the continents, and much sedimentary debris from the lands covers the oceanic basalt close to the continents. The substances released from sea water or embodied in the skeletons of minute organisms accumulate slowly over much of the ocean floor.

Below the lower boundary of the Earth's crust is a very thick shell of another dark rock which differs from basalt in its chemical composition and mode of origin. This shell is the *mantle*, about 1800 miles (2900 kilometers) thick. The lower boundary of the lithosphere lies within the mantle at a depth which ranges from about 45 to 60 miles (70 to 95 kilometers). The level of this lower boundary is determined by temperature: above the boundary, the rocks are rigid and relatively brittle; below the boundary, in a zone of variable thickness, the rocks are exceptionally hot and relatively plastic. This plastic zone is the *asthenosphere*. The enormous, semi-rigid plates into which the lithosphere is divided move about slowly and ponderously as though floating on the asthenosphere. In some places, parts of the mantle have moved upward along great fractures in the crust and have been exposed at the surface. The thickness of the mantle and the presence of the plastic zone is revealed by the behavior of earthquake waves that pass through the interior of the Earth on their way to distant receiving stations. The velocity of these waves changes as they cross boundaries between different materials or different temperatures. Some waves are retarded in hot plastic materials and completely absorbed in molten materials.

11

The lateral boundaries of the half dozen or so great plates of the Earth's lithosphere are of three principal types. In the first type, the plates are moving away from each other. The zone of separation is marked by fractures which are periodically filled with basaltic lava from below. Later, the filling itself may be split by renewed separation and the new fractures again occupied by basalt. Some of the basalt spills out as lava flows on the ocean floor. The second boundary type is found where plates are moving laterally past each other, as along the San Andreas Fault in coastal California. The third boundary is an overlapping type where plates are converging. In this case, the edge of one plate dips under the other and is progressively absorbed at depth. This descent of one plate under another is known as subduction. In all three types, relative plate motions amount to only 0.5 to 6.0 inches (1–15 centimeters) per year.

Plate boundaries are the sites of much volcanic activity. At boundaries where plates are separating, as at the Mid-Atlantic Ridge and the East Pacific Rise, lava rises from the depths by way of the great fractures. Submarine volcanoes and lava flows are common at such sites. In Iceland, the Mid-Atlantic Ridge has actually been raised above sea level, providing opportunity for direct observation of the associated volcanic activity. At boundaries where an oceanic plate dips under a continental plate, the descending edge becomes heated up in the depths and molten material works its way to the surface inland from the continental border. Thus, most volcanoes are located either along oceanic rifts or where oceanic plates descend under the continents. The volcanic "Ring of Fire" around the Pacific Ocean marks the sites of such descending plates. A few other volcanoes, however, such as those of Hawaii, are in the interior of plates and have a different origin which we need not go into.

The lateral movement of crustal plates may be readily observed along great fractures such as the San Andreas Rift. Here the oceanic plate is moving northwestward relative to the continental plate. Although the average rate of movement is small, two inches (5 centimeters) or less per year, the shift in any one event may be considerable. Thus, in the San Francisco earthquake of 1906, the ground locally shifted as much as 21 feet (6.5 meters), displacing roads, fences, pipelines, and other structures.

The characteristics of spreading oceanic rifts have been studied not only on land in Iceland, but by direct observations from deep-sea submersibles (Photo 6). Where spreading is slow, an oceanic ridge of appreciable height may be built up by the outflowing lava. The Mid-Atlantic Ridge is an example. Actually, such ridges are extremely broad welts with very gentle slopes. When the lava ceases to rise, continued spreading opens up a central trench along the ridge. The rates of movement have been determined by noting the distances to which progressively older basalts have been carried away from the ridges. Ages of the basalts are determined by measurements of radioactive disintegration of certain minerals in samples collected from the sea floor. We shall have more to say about this method of dating later.

Figure 3 is a generalized diagram of the eastern Pacific illustrating seafloor spreading and two types of plate boundaries. In the front face of the diagram, lava is shown reaching the ocean floor by way of a fracture through an oceanic ridge — in this case, the East Pacific Rise (EPR). The lava is carried laterally in both directions (small arrows flanking the rise) on the moving lithosphere, which includes the oceanic crust (OC). The spreading, which varies from about one-half to four inches (1 to 10 centimeters) a year, reopens the fracture, permitting more lava to reach the sea floor.

The sea floor is divided into parallel segments by large cross fractures or *transform faults* (TF) at right angles to the ridges. Along these faults the plate segments move differentially relative to each other. On approaching the continental margin, the sea floor descends to create a deep trench into which sediments from the land are deposited. As motion continues, these sediments are crumpled up and part is carried down below the edge of the continent. The descending plate partially melts at depth, and the resulting molten rock or magma escapes to the surface to create volcanic activity. An oceanic ridge becomes inactive when it descends under the edge of the continent and the lava

13

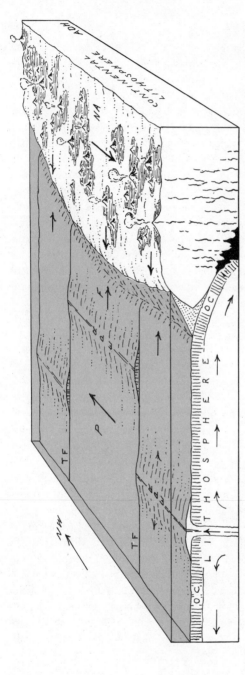

Figure 3. Generalized diagram illustrating seafloor spreading and an overlapping plate boundary. Spreading occurs at oceanic ridges. Subduction is the descent of one plate under another. This view suggests conditions along the California coast many millions of years ago (NA—North American Plate; F—Farallon Plate; P—Pacific Plate; EPR—East Pacific Rise; OC—Oceanic Crust; TF—Transform fault.) The dashed arrows on either side of the East Pacific Rise indicate spreading lava. The small solid arrows on the ocean floor east of the rise indicate the direction of motion of the ocean floor toward the continent. The largest arrows show the relative global motions of the North American and Pacific plates. Although the segments of the ocean floor east of the rise are moving landward, the entire ocean floor assemblage is moving north.

outlet is sealed off. The portion of the North American Plate (NA) shown in the diagram represents the ancient coastal region of California from the sea to the Sierra Nevada. Between the continent and the East Pacific Rise is the Farallon Plate (F). West of the rise is the Pacific Plate (P). The large arrows indicate the gross relative movements of the ocean floor and North American Plate. The moving sea floor bodily carries the complex of ridges and transform faults to the northwest.

By backtracking the movements of the great continental plates, it appears that they were once part of a single master continent, *Pangaea* (Pan-jee-uh), about 200 million years ago. Pangaea began to split up about that time, and the various segments have since been separating, leaving new ocean floor in the widening gaps.

Because of the inexorable forces in operation at plate boundaries, the boundaries are the sites of mountain-making, large-scale faulting, earthquakes, and volcanic activity. Coastal California lies in one of these unstable belts, and its landscape shows the effects of these plate movements. The spreading floor of the eastern Pacific Ocean has profoundly affected the mid-California coast, at times crumpling it into mountains. The development of the San Andreas Rift and other great faults has made the area earthquake-prone. It is appropriate at this time to consider the cause of earthquakes, a major hazard in California.

Let us imagine an area crossed by a road trending east and west as in Figure 4. Let us suppose that a fault (F–F) crosses this area. Let us further suppose that this area is subject to forces acting in the directions of the arrows. The left side of the block will move north, the right side south, and the rocks in the central area will be flexed. The flexing will extend over a broad zone and will distort the road as shown in position 2 of

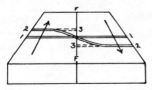

Figure 4. Elastic rebound theory of earthquakes.

15

the diagram. Eventually, the warping forces may overcome friction along the fault and a new displacement will occur. The severed ends of the road, no longer tied to each other, spring back elastically to positions (3–3). The snapping of the rock results in vibrations which are propagated outward as earthquake waves.

The faulting temporarily relieves the strain on the rocks, but the responsible forces still remain and continue to deform the rock. The fault surface is not a smooth plane along which sliding can take place continuously. On the contrary, it is very irregular and offers considerable resistance to further movement. In spite of its presence, therefore, the rocks once again begin to bend and continue to do so until the limit of frictional resistance along the fault is reached. A new sudden movement then takes place. These repeated movements generally occur at irregular intervals, making earthquake prediction very difficult. To add to the difficulty, there are so many faults in the Coast Ranges of Middle California (Figure 5) that the strains set up in the rock are sometimes relieved by displacement on one fault and sometimes on another. As yet, no precise way has been found to predict on which fault the next movement will take place or even where, on a single fault, the site of the next movement will be, but some progress is being made. Most of the faults, like the San Andreas Fault, are characterized by horizontal shifting of the ground.

Let us now inquire as to how a geologist can identify areas of past volcanic activity and locate the sites of long-vanished volcanoes. The lavas that poured out on the surface in ancient times, and the ash and cinders (pyroclastic materials) that were exploded into the air, were no different from those of today. Thus, recognition of areas of past volcanic activity is merely a matter of being able to recognize ancient lavas and explosion products. The location of the volcanoes themselves is another matter. The lava that is ejected from volcanoes comes from great depth. The volcano itself is simply a cone of volcanic rocks built up at the surface end of a long pipe leading from the depths (Figure 6). Sometimes during the life of a volcano,

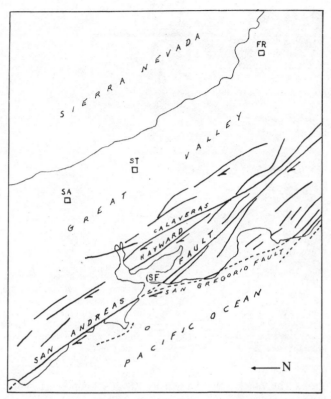

Figure 5. Active and possibly active faults in the Coast Ranges of Middle California. (After T. W. Dibblee, Jr.)

explosions will crack the sides of the cone; the cracks formed will radiate outward from the center. Later, molten rock from the depths may squeeze its way into these fractures and solidify, forming features known as *dikes*. When the volcano becomes extinct and its flanks are deeply eroded, the rocks in these fractures may stand up like radiating ribs, and the central pipe will rear upward as a great monolith, a huge isolated spire known as a *volcanic neck* or *plug*. Thus, the central monolith and the radiating ribs constitute the skeleton of the volcano. There are several examples of volcanic necks in the Bay Re-

17

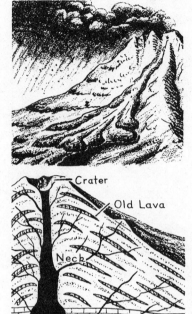

Crater

Old Lava

Neck

Figure 6. Sketch and section of volcano. The section shows the magma reservoir in depth, the neck leading to the surface, and the dikes (thin lines) that fed lava flows at different stages of volcanic growth.

gion. The black spire known as Lover's Leap, near which Highway 152 passes just east of Bell Station on its way to Pacheco Pass, is a fine example. Ancient lavas are also widespread in the Bay Region. One is exposed in the quarries back of Stanford University, where the rock was once excavated for use in road building. The entire crest of the Sonoma Range consists of old lava flows and pyroclastic deposits.

We know now how the geological detective determines the geography and topography of the past, and how he locates centers of volcanic activity. Let us inquire next how he deciphers ancient climates. One easy way is by studying the fossils entombed in the sediments. We know, for example, that the different climatic zones of the Earth are characterized by different assemblages of plants and animals. When we find fossil palms and the remains of alligators in a sedimentary formation, we are justified in assuming that the sediments in which they are entombed were laid down in a warm, moist, subtropical climate, for such is the climatic environment of these organisms today. In contrast, fossils of redwood, elm, ash, walnut, and similar trees indicate a moist temperate climate. And the remains of the woolly rhinoceros and woolly mammoth indicate glacial climates, even in regions that now are warm. Ancient climates are often indicated, too, by the composition of the sediments and certain inherent structures. The presence of salt layers, for example, commonly indicates intensive evaporation, presumably in a dry climate. Dry climates are also indicated by the wind-rounded shapes and frosted appearances of the sand grains in certain sandstone formations, and by the presence of a peculiar curved layering known as *eolian cross-bedding*.

As for dating the past, the geological detective now relies largely on methods developed during the atomic age— methods based on the phenomenon known as radioactivity. A radioactive element is one that loses part of its substance by radiation and changes into progressively lighter elements at a steady, unchangeable rate. Uranium, for example, after changing from one element to another, finally ends up as nonradioactive lead. One of the elements in the uranium-to-lead sequence is radium, a substance that is used in painting luminous watch dials. The actual disintegration of radium atoms may be observed in a darkened room by viewing the luminous hands or numbers of a watch or clock under a strong magnifying lens. The flashes of light are miniature explosions, each one representing the disintegration of one radium atom. The time for half of a given amount of uranium to pass through the complete sequence of changes to lead is very long. Since the rate of change is known, one can determine how long ago a uranium mineral came into existence by comparing the amount of the remaining uranium with the amount of surrounding lead into which the missing uranium disintegrated. This method has been used successfully in dating rocks more than 3 billion years old. The uranium-lead method, however, is useful only for dating very distant events involving many millions of years. Other radioactive elements such as potassium and carbon are used for shorter spans of time. Radioactive potassium is accurate for dating events between 500,000 and 100 million years ago. The granite of Montara Mountain on the San Francisco Peninsula and the granite of the Farallon Islands have been shown by the potassium method to be about 90 million years old, whereas the granites of Pt. Reyes and the Santa Lucia Range are only 84 and 82 million years old respectively. Volcanic ash is particularly valuable for dating because it settles more or less instantaneously over large areas. Thus, if the age of the ash can be determined, it provides a precise date for the deposits in which it occurs and permits correlation of events over wide areas. The radioactive carbon method is useful for dating within the last 40,000 years. By measurement of the radioactive carbon in fossil seashells, it is now known that the lowest of the marine terraces along the coast of Middle California was formed more than 40,000 years ago, when sea level stood much higher than now. Inasmuch as other evidence indicates that the last high stand of the sea was about 100,000 years ago, it is believed that the terrace is probably that old.

Another method for dating and correlation is based on the recently discovered fact that the positive and negative magnetic poles of the earth have reversed repeatedly in the past. Thus, at times, magnetic compasses pointed south rather than north. The significance of this to our geologic detective is as follows. Many lavas contain grains of the naturally magnetic mineral, *magnetite*. During cooling of the lava, each of these grains orients itself with the earth's magnetic field like a little magnet. When the lava solidifies, this direction of ancient magnetism (*paleomagnetism*) is locked in the rock. Means are available to determine the orientation of this paleomagnetism. Such determinations have been made for a great number of lavas of different ages all over the world, and a time scale recording the reversals in magnetic polarity has been prepared for the last few million years. By identifying the nature of the polarity in ancient lavas, it is possible to date ancient events and correlate them over long distances. It has also been found that magnetic minerals settling out of water will orient themselves magnetically; hence ancient sediments are also proving useful in paleomagnetic dating and correlation.

With this background into the clues used by the geological detective, we may proceed to our story of the landscape of Middle California.

THE PROBLEM OF THE BEGINNING

Where shall our story begin? Latest estimates place the age of the Earth at approximately 4.5 billion years. Geologists have divided this vast span of time into subdivisions, much as the year is divided into months, weeks, and days. The major subdivisions of this geologic time scale or calendar are shown in the accompanying table. Note that the first 4 billion years of Earth history are referred to as Precambrian time. So little is known of these ancient rocks that subdivision is practically impossible. There are several reasons why these rocks provide little information about the Earth's infancy and adolescence. One is that over much of the Earth these rocks are deeply buried under other rocks and are not visible to us. We see them generally where great canyons like the Grand Canyon of the Colorado River have eroded through the covering rocks, or where the ancient rocks have been punched up in the cores of mountain ranges and exposed by erosion. Secondly, these ancient rocks have been crushed and broken so many times during the long and turbulent history of the Earth that their original characteristics have long since been obliterated. Hence they tell us little of the conditions at the time of their origin. These ancient rocks, too, were formed when life was exceedingly scarce on the Earth. The almost complete absence of fossils deprives us of some of our best clues to ancient environments.

The past 570 million years of Earth history are much better known, because many of these rocks have been changed very little from their original condition. It has not only been possible to divide the past 570 million years into major subdivisions called eras, but even to subdivide the eras into smaller divisions known as periods, and these in turn into smaller units known as epochs.

The hills and mountains of Middle California are carved out of rocks of many different ages. Rocks of undoubted Precambrian age, however, are unknown. The oldest rocks in the region are probably of late Paleozoic age. These are widely exposed in the Santa Lucia Range but occur only as small patches in the Gabilan and Santa Cruz ranges. The remnants are so small and so badly altered by the ravages of time that they tell us little about the conditions that prevailed when they were deposited. Furthermore, since these ranges lie west of the San Andreas Fault, they are not "native" to this area; they moved into Middle California only about 20 million years ago. Prior to their arrival, there was only sea west of the San Andreas Fault. We shall attempt later to describe the evolution of these ranges during their migration north.

Rocks of Mesozoic age, formed when dinosaurs ruled the Earth, underlie large parts of the San Francisco Bay region from the ocean to the Great Valley and reappear in the Sierran foothills. They are the foundation rocks of much of San Francisco, of the eastern foothills of the Santa Cruz Range, and of large areas in the Diablo Range. North of the bay, these rocks extend from the Golden Gate northward into Oregon, and from the Pacific Ocean to the Sacramento Valley and the Sierran foothills. Although the rocks are largely sedimentary, they include some igneous rocks. Some of these are dark; others are light. The lighter ones are granite, the same kind of rock in which Yosemite Valley is carved. In the Bay Region, the granites are exposed in Inverness Ridge, Pt. Reyes, the Farallon Islands, Montara Mountain, Ben Lomond Mountain, and the Gabilan and Santa Lucia ranges. The Mesozoic and early Tertiary rocks west of the San Andreas Fault also came into this area from far to the south; but being younger, they have not traveled as far as the Paleozoic rocks.

The early Tertiary rocks are not as abundant as the Mesozoic rocks and are more widely scattered throughout the Bay Region. They form some of the ridges that encircle Mt. Diablo (see Photo 3). The thinly banded rocks in the vicinity of the Broadway Tunnel in Oakland and at San Pedro Point on the Coast Highway south of San Francisco are also early Tertiary in age. Additional examples may be seen at Pt. Reyes and at Point Lobos south of Carmel.

Both the Mesozoic and early Tertiary rocks contribute to the

THE GEOLOGIC CALENDAR

SELECTED LOCALITIES

ERA	PERIOD	EPOCH	SELECTED LOCALITIES
Cenozoic	Quaternary	Holocene (Recent)	Sand of Ocean Beach and on hills of western San Francisco. Alluvium of river bottoms. Silts and muds of Sacramento delta.
		10,000 B.P.*	
		Pleistocene	Terraces near Millerton: Sand and gravel. Road cuts at Colma: Old beach and dune sands. Quarries at Irvington: Sand and gravel. Volcanic rocks of Mt. Konocti at Clear Lake.
		1,800,000 B.P.	
		Pliocene	Road cuts between Tomales and Dillon Beach: Sandstone. Road cuts in upper Claremont Canyon: Sandstone, shale, conglomerate. Bald Peak and Grizzly Peak: Basalt. Little Grizzly Peak: Rhyolite breccia. Road cuts between Rodeo and Oleum: Tuff. Coast south of Half Moon Bay: Black shale.
		5,000,000 B.P.	
		Miocene	Lower part Claremont Canyon: Chert, limestone. Quarry on west side Inverness Ridge on Point Reyes road: Chert, shale. La Honda area: Basalt. Natural Bridges State Park, Santa Cruz: Shale. Road cuts in Monterey town: Sandstone, shale. Pinnacles National Monument: Rhyolite volcanic rocks.
		23,000,000 B.P.	
	Tertiary	Oligocene	Road cuts on Route 9 between Saratoga and Santa Cruz at Riverside Grove: Sandstone and shale. Road cuts along Route 17 about 5 miles south of Los Gatos: Sandstone and shale.
		38,000,000 B.P.	
		Eocene	Road cuts south of Antioch Reservoir: Sandstone and shale. Road cuts in vicinity of Woodside: Sandstone and shale. Early gold-bearing gravels in the Sierra Nevada.
		55,000,000 B.P.	
		Paleocene	Road cuts west of Martinez: Sandstone and shale. Coast Highway between San Pedro Point and Devil's Slide: Shale and sandstone. Conglomerate, sandstone, shale at Pt. Reyes and Pt. Lobos.
		65,000,000 B.P.	
	Cretaceous		Road cuts along Route 28 in Vaca Mountains: Sandstone, shale, conglomerate. Road cuts in Niles Canyon: Sandstone and shale. Coast Highway between Devil's Slide and Moss Beach: Granite. Inverness Ridge: Granite.
		140,000,000 B.P.	

Era	Period	Age	Examples
Mesozoic	Jurassic	195,000,000 B.P.	Almost all road cuts in San Francisco: Sandstone, shale, chert, dark igneous rock, serpentine. Road cuts north of Golden Gate and in Mt. Tamalpais State Park: Sandstone, shale, chert, basalt. Skyline Drive from Milbrae turnoff south to Woodside: Sandstone, shale, dark igneous rock, serpentine. Mariposa slates near Mariposa in the Sierra Nevada.
	Triassic	230,000,000 B.P.	
Paleozoic	Permian	280,000,000 B.P.	
	Pennsylvanian	320,000,000 B.P.	Bear Valley Ranch in Inverness Ridge: Quarry. White limestone. Road cuts 12 miles south of Carmel along Highway 1: White limestone. Road cuts between Big Sur and Lucia along Highway 1: Mica-rich metamorphic rocks. Metamorphic rocks of Calaveras Formation at Geologic Exhibit along Yosemite Highway 6 miles east of Briceburg.
	Mississippian	345,000,000 B.P.	
	Devonian	395,000,000 B.P.	
	Silurian	435,000,000 B.P.	
	Ordovician	500,000,000 B.P.	
	Cambrian	570,000,000 B.P.	
	Precambrian	4,500,000,000 B.P.	

*B.P.: Before Present.

details of the landscape because of their variable resistance to weathering and erosion, but they are not responsible for the gross forms of the modern ranges. During the long span of time when these rocks accumulated, some under the sea, some on dry land, some by volcanic eruption, and some by molten injection below the surface, generations of mountain ranges were upheaved, only to be later worn to their roots. The topography that existed when these older rocks were formed has long since disappeared.

Where in this repetitious sequence of events shall we start our story? Because the Paleozoic rocks of the Sierran foothills are "native" to Middle California, we will start our story when they were first formed several hundred million years ago. The Coast Ranges and Great Valley did not then exist. This early history is not well understood; hence our discussion will necessarily be cursory. At about 150 million years B.P., the story becomes clearer and we are able to go into more detail. Although 150 million years is an enormous span of time, it amounts to only one-thirtieth of the long span of Earth history. By way of illustration, if we were to assume that the height of one of the towers of the Golden Gate Bridge, 750 feet (229 meters), represents the 4.5 billion years of Earth history, the upper 25 feet (7.6 meters) would represent the entire 150 million years, and the thickness of a dime atop the tower would represent more than all of recorded human history.

Now we face another problem: how to simplify a story that is exceedingly complex. Our story would have been simple if each event in the past, such as invasion by the sea, upheaval of mountains, and volcanic eruption, had affected the entire region simultaneously rather than piecemeal. We would merely have to catalogue the successive events in chronologic order. But each event affected only part of the region. While some localities were being invaded by the sea, others were being crumpled into mountains or being buried under sedimentary or volcanic accumulations. In other words, each separate area had a history of its own which may have been entirely unrelated to the history of the immediately adjacent areas. Even if we knew all the details of the complex history of Middle California, which we do not, any attempt to describe every small episode in the creation of the present landscape during the past 150 million years would involve us in such a welter of detail that the essential elements of the story would be lost. For the sake of clarity, then, we shall select only the major episodes in the geologic history. To simplify the story still further, we shall telescope events; that is, we shall consider as having happened simultaneously, events that were actually separated in time. To help us visualize the major events more clearly, a sequence of diagrams is presented. Each landscape represents a composite of the major landscape features that prevailed during the indicated period of geologic time. And so, to our story.

THE DISTANT PAST

MIDDLE PALEOZOIC TO EARLY CRETACEOUS TIME

During the second half of the Paleozoic Era, that is, up to about 230 million years ago, much of middle California lay beneath the sea. There was only one vast ocean on the globe at that time, the Pacific, surrounding a single master continent. It was the breakup of this continent that resulted in all the smaller continents of today. On the sea floor where the Sierra now lies, there accumulated a thick assemblage of marine shales, siltstones, sandstones, limestones, and volcanic deposits. At the close of the Paleozoic and early in the Mesozoic Era, these rocks were crumpled by subduction (see page 12) to create an ancestral Sierra Nevada. The sediments were metamorphosed to slates, phyllites, cherts, quartzites, marbles, schists, and gneisses, and these were invaded by granitic magmas in Triassic time. These granites, the oldest in the Sierra, lie east of our area of interest.

The Paleozoic metamorphic rocks of the ancestral Sierra Nevada are collectively known as the Calaveras Formation. They are exposed today in the Sierran foothills along the Yosemite Highway (California Highway 140) at Briceburg and almost continuously along the Merced River for the next dozen miles or so to the east. At the Geologic Exhibit, about 6 miles (10 kilometers) east of Briceburg, tightly folded quartzites and cherts of this formation are prominently exposed (Photo 5). The Pacific shoreline at the time of these ancestral mountains was probably within the present foothills.

There then ensued, during the Triassic, Jurassic, and much of the Cretaceous periods, an exceedingly complex and poorly understood sequence of events. Periodic subduction revived and accentuated the mountain-building forces, and additional granitic magmas invaded the crust. The deformation seems to have reached a climax in the late Jurassic period. At this time, the Triassic and Jurassic seafloor sediments, as well as the volcanic rocks of an island arc that lay offshore, were mashed against the ancient coast in a mountain-making episode known as the Nevadan Orogeny. The sediments and volcanic rocks were altered to a second series of metamorphic rocks and shoved against the older Calaveras rocks. One of the widespread members of this later series is the Mariposa Slate, which, together with its associated metamorphic rocks, may be seen along the Yosemite Highway for 6 to 8 miles (10 to 13 kilometers) east of the Merced County line. The slates commonly project above the foothill slopes like tombstones (Photo 8). In addition to the granites that were generated deep below the surface, long lenticular bodies of dark igneous rock, now altered to greenish serpentine, were tectonically raised into the metamorphic complex. Thanks to subsequent deep erosion, the serpentines may be seen today on both sides of the town of Mariposa on the Yosemite Highway. More extensive exposures appear along the Mother Lode Highway (California Highway 49), starting at the community of Bear Valley about 15 miles (24 kilometers) north of Mariposa and continuing almost uninterruptedly for the next 6 miles (10 kilometers) north. These

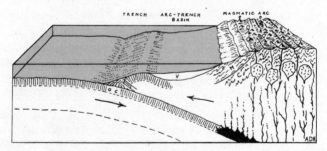

Figure 7. Latest Jurassic–early Cretaceous landscape, 150 to 100 million years B.P. In the Sierra, the Mariposa assemblage of rocks (M) was mashed against the Calaveras rocks (C) during the Nevadan Orogeny. Note the new trench formed by subduction offshore, the buckled edge of the upper plate, and the basin between it and the mainland. Melting in depth of the leading edge of the descending plate resulted in volcanic activity at the surface. Granite batholiths formed several miles below the surface. Erosion of the magmatic arc added much volcanic debris (V) to the early Cretaceous sediments of the arc-trench basin.

29

rocks commonly form bold ridges in the present Sierran foothills.

Because of the uncertainty surrounding the sequence of events during the late Paleozoic and early Mesozoic history of the Sierra, we can say little more about these ancient events. We begin our somewhat more detailed story in the latest Jurassic and early Cretaceous periods, when the future sites of the Coast Ranges and Great Valley were delineated.

After the Nevadan Orogeny, the site of subduction shifted from the Sierran foothills 80 miles (130 kilometers) seaward to the present site of the Coast Ranges. Figure 7 is a synthesis of some of the events of latest Jurassic and early Cretaceous time, 150 to 100 million years ago. On the Sierran mainland, we see the Jurassic Mariposa assemblage of rocks (M) mashed against the older Calaveras rocks (C). Far offshore is the new subduction zone with a trench at the site of the descending plate of oceanic lithosphere. With continued subduction, the seaward edge of the upper plate is buckled, or rises isostatically, to form a submarine ridge and a broad arc-trench basin landward (Figure 7). A modern example of this stage of plate tectonics is provided off the south coast of Java, where a submarine ridge, with its crest 1–1.5 miles (1.6–2.4 kilometers) below sea level, lies seaward of a basin more than 3 miles (5 kilometers) deep. The arc-trench basin off the ancient California coast trapped large quantities of sediment both from the land mass to the east and from the ancestral Klamath Mountains to the north, outside our area of interest. The sediment was carried west and south by heavily-laden, bottom-hugging currents known as turbidity currents.

By early Cretaceous time, the leading edge of the descending oceanic plate had reached depths under the continent sufficient to cause partial melting. The resulting magma rose to the surface, resulting in widespread volcanic activity and the creation of a magmatic arc. The volcanic rocks formed at this time have long since been removed by erosion, but the debris is preserved in the sediments of the Great Valley. These volcanic-rich sediments are indicated by a V in Figure 7. Additional granite batholiths, huge bodies of molten rock, were generated at this time and added to the collection of batholiths formed during the Triassic and Jurassic periods.

Part of the sediment washed into the California arc-trench basin spilled over into the trench, but part of the trench sediment probably resulted from landslides and turbidity currents generated on the slopes of the trench itself. Such were the beginnings of the so-called Franciscan assemblage of rocks so abundantly exposed in San Francisco and its environs.

The trench and basin sediments, except close to the ancient landmass, were largely deep-water types. The evidence for this is the almost complete absence of fossils with calcium carbonate shells such as are found along the shore and in shallow water today. Calcium carbonate dissolves at depths greater than 10,000 to 15,000 feet (3,000 to 4,500 meters), and hence is absent in deep-sea sediments. The few shallow-water fossils that have been found are believed to have been transported from sites nearer land by far-reaching turbidity currents. These early Cretaceous sedimentary deposits in the arc-trench basin are part of what is known to geologists as the Great Valley Sequence, possibly as much as 40,000 feet (12,000 meters) thick and including strata from Late Jurassic to Late Cretaceous in age.

Late Cretaceous Time
100 to 65 million years B.P.

By late Cretaceous time, as a result of continued elevation, erosion in the Sierra had removed the cover of volcanic rocks and considerable thicknesses of the underlying metamorphic rocks (Figure 8). Thus, large areas of the formerly deeply buried granites, including the Cretaceous granites that provided the gold of the Mother Lode quartz veins, were now exposed. Erosion of these granites released a distinctive mineral, orthoclase, to the late Cretaceous sediments (Figure 8, G) that were spread into the Great Valley. These orthoclase-rich sediments may be observed today in roadcuts along California Highway 128 about 2.8 miles (4.5 kilometers) southwest of Monticello Dam on Lake Berryessa due west of Sacramento. We see also in Figure 8 that continued subduction has intensified crumpling of the Franciscan sediments (F) accumulating

31

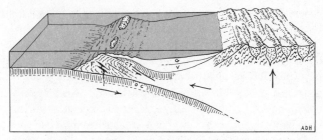

Figure 8. Late Cretaceous landscape, 100 to 65 million years B.P. Elevation of the Sierra resulted in accelerated stream erosion and removal of the volcanic cover and much of the metamorphic rock from above the granite batholiths. Considerable granitic debris (G) was added to the sediments in the offshore basin. In the trench, the accumulating sediments were crumpled and sheared and parts were forced under the edge of the basin sediments along the Coast Range Thrust (CT). The trench sediments comprise the present Franciscan Assemblage (F). The buckled edge of the upper plate appeared locally above water, probably as a string of islands. (OC—Oceanic crust.)

in the offshore trench. Because deposition continued during crumpling, slabs of younger and younger sediment were wedged under the buckled upper plate. The wedging elevated the trench deposits to ever greater heights; but except for scattered islands, this ridge was to remain a largely submarine feature until at least early Tertiary time.

The westernmost portions of the Great Valley Sequence (Figure 8, G,V) were not immune to the forces that were crumpling the Franciscan rocks to the west. They, too, were locally folded, but the folds were generally shallow and died out rapidly toward the east. At their extreme western edge, however, the relations were more complicated. Subduction of the oceanic plate forced portions of the Franciscan rocks under the Great Valley Sequence, the boundary being an extensive, low-angle, eastward-dipping fault. This is the Coast Range Thrust (Figure 8, CT), traceable for many miles up and down the Coast Ranges. This underthrusting of Franciscan rocks continued, with one long interruption, for millions of years.

Some geologists believe that horizontal ground-shifting along an ancient San Andreas Fault System may have reached middle California about 75 million years B.P. and expanded north during the next 15 million years or so. This ancient fault system is believed to have become activated when subduction of an oceanic plate (Kula plate) older than the Farallon plate, ceased in this area and the edge of the ancient continent was torn off by the northward movement of the Pacific floor. It is as yet unknown where this proto–San Andreas Fault System was located. In any event, as the moving ocean floor carried the still actively subducting Farallon plate into middle California, activity along the proto–San Andreas Fault System ceased. Only when subduction of the Farallon plate terminated in turn did the present San Andreas Fault System become active. This second stage of activity, began much later in our story.

The central areas of the Great Valley were also affected by the deforming forces that were crumpling the Franciscan rocks and folding the western members of the Great Valley Sequence. The valley floor was differentially warped with deepest subsidence in the Sacramento Valley north of Sacramento and in the Coalinga region of the San Joaquin Valley. In these subsiding areas, enormous thicknesses of Cretaceous sediments were deposited. Between these basins there remained a broad residual submarine arch extending west from the Sierran foothills under the present site of the city of Stockton. The arch itself continued to rise and was destined, in later time, to rise above sea level.

PALEOCENE TO EARLY MIOCENE TIME

65 to 21 million years B.P.

The activation of the modern San Andreas Fault System was an important event in the Cenozoic history of California. Movement within the fault system began in southern California when the East Pacific Rise, separating the Pacific and Farallon plates (see Figure 3), reached the continental border about 29 million years ago. The rise entered the subduction zone, and subduction ceased because the northward-moving Pacific plate was now against the continent and its motion was more or less parallel to the coast. The Pacific plate began to carry along with it a long coastal sliver of the southern California borderland, now largely submerged. This movement probably took place along a fault or faults near the edge of the continental shelf. As time went on, the movement shifted to more easterly faults of a developing fault system, but not in any clear progressive order. The present appearance of the above-water portions of the moving borderland probably bears little resemblance to their initial configurations. The impact of the modern San Andreas Fault, as distinguished from possible precursors, probably did not reach Middle California until about 21 million years ago, taking over as other faults became inactive.

Much of the evidence for cessation of subduction and the initiation of movement along the modern San Andreas Fault comes from terrain west of the fault, including the adjacent sea floor. But this seaward belt of country is foreign to Middle California; it has been migrating northward for millions of years and has only recently reached its present positon. The evolution of these coastal topographic features began in southern California and has been influenced by geologic events that took place during the migration north. Thus, the record that we read today in the terrain and sea floor west of the San Andreas

Fault in Middle California is actually a record acquired during the long migration north.

Because most details of the complex migrational history of the California borderland are still unknown, we will avoid these complications and consider all movement as having taken place along a single San Andreas Fault. Even this, however, requires some clarification. Normally, when we speak of the San Andreas Fault, we are referring to the latest fracture along which movement has taken place. But during the past, slippage has occurred along many closely spaced parallel faults within a zone from less than half a mile to several miles wide. This is the San Andreas Fault Zone. The repeated slicing of the earth's crust within this zone has churned and broken up the rocks, making the zone very susceptible to stream erosion. This explains why the fault zone forms long straight valleys for considerable parts of its length. But the San Andreas Fault is only one member, although presently the most important one, of a whole family of roughly parallel faults, some of which are many miles apart. Together these constitute the San Andreas Fault System. Other members of the fault system in Middle California include the Hayward and Calaveras faults, the long San Gregorio Fault, and many other roughly parallel fractures (see Figure 5). When we speak in this book of northward movement of the terrain west of the San Andreas Fault, we do not mean to imply that this entire coastal segment moved as a unit. At different time, slippage was along different members of the fault system. Because of this differential slicing, the present appearance of the ranges and valleys that have moved northward into Middle California may have little resemblance to their appearance before migration. It is probable that the first horizontal movement took place along a member of the fault system that now lies offshore at the edge of the continental shelf.

In spite of the above difficulties, we do have some general information bearing on the history of the topographic features that eventually moved into our area—that is, the Gabilan and Santa Lucia ranges, the Santa Cruz Range, the Point Reyes

35

Peninsula, and the long sliver terminating northward at Point Arena. On the other hand, we know nothing of the topographic features that may once have been native to this part of the coast but have since moved northward out of our area.

But to return to our story. We noted that at the close of the Cretaceous Period there was a long period of erosion during which the folded Cretaceous strata (Great Valley Sequence) on the west side of the Great Valley were eroded across. The first of the Tertiary sediments rest on this erosion surface and locally lap over even onto Franciscan and granitic rocks.

Within the Great Valley, the early Tertiary seas fluctuated widely because of differential upwarping and subsidence of the valley floor. Figure 9 shows the maximum composite extent of the seas during this 44-million-year interval. A seaway may have connected the open sea with the Great Valley. This is suggested by the thickening of the early Tertiary marine sediments of the Great Valley toward this gap and the convergence of buried canyons, which we shall describe shortly, in that direction. For convenience, we shall refer to this seaway as Markley Strait, after the name of one of the canyons. There may have been other connections between the Great Valley and the sea, however, as suggested by the presence of early Tertiary marine sediments elsewhere within the Coast Ranges. It is possible that the Coast Ranges consisted of a complex of islands at this time. The presence of early Tertiary marine sediments just south of Clear Lake (Figure 9, L) suggests that an embayment from Markley Strait penetrated at least that far north of the present bay lowland.

Early Tertiary sediments were also deposited on the ocean side of the ancestral Coast Ranges, where they, too, were crumpled and faulted by continuing subduction. Parts were raised above sea level and plastered against the seaward side of the largely submerged ridges, expanding the accumulating pile seaward. These incipient Coast Ranges were probably differentially uplifted, although direct evidence for this has probably long since moved northward out of Middle California. At several places west of the San Andreas Fault, however, as at San Pedro Point, Point Lobos, and Point Reyes, the early Tertiary

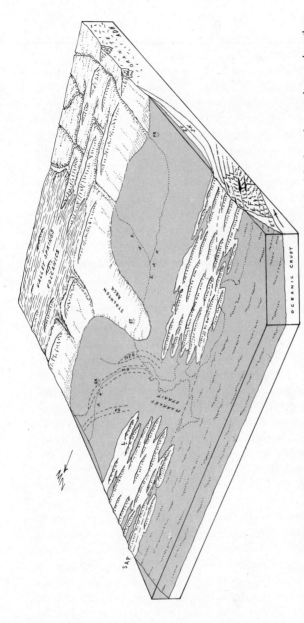

Figure 9. Early Tertiary landscape, 65 to 21 million years B.P. The Stockton Arch separated the inland sea into northern and southern segments. The inland sea was connected to the open ocean by the Markley Strait. Three buried gorges are indicated by dashed lines. None of the modern landscape features west of the San Andreas Fault had yet reached Middle California. In the east, the northern Sierras are shown blanketed by the Valley Springs volcanics, which largely obliterated the older drainage. (ET—Early Tertiary sediments; F—Franciscan Assemblage; FR—Fresno; J–K—Jurassic-Cretaceous sediments; L—Lower Lake; MEG—Meganos Gorge; MG—Markley Gorge; MR—Merced River; PG—Princeton Gorge; SA—Sacramento; SAF—San Andreas Fault; SJR—San Joaquin River; SR—Sacramento River; ST—Stockton.)

marine sediments have a pebble conglomerate or a bouldery accumulation at their base, which suggest coastal conditions (Photo 9). Although these sites were south of our area at the time of accumulation of these deposits, comparable conditions may have prevailed in Middle California. At least we know that swampy conditions existed in the Mt. Diablo area, because coal beds are present in the Eocene deposits there.

During much of early Tertiary time the Great Valley was separated into two major basins by a rising upwarp of the sea floor. This upwarp, known as the Stockton Arch, had been a submarine feature in the late Cretaceous. At times, the seas receded from large parts of the interior and deep canyons were excavated in the exposed sea floor, probably by rivers from the Sierra Nevada crossing the plain. Seaward of the ancient shorelines, however, the canyons were probably eroded by submarine turbidity currents. The existence of these now deeply buried valleys has been revealed by deep drilling for oil and gas.

The oldest of the known gorges shown in Figure 9 is the Meganos Gorge (MEG). This gorge, of late Paleocene age, has been traced for about 45 miles (72 kilometers) from its head near Walnut Grove to the Mount Diablo area, where subsequent uplift has exposed part of its sedimentary fill. Total depth of the gorge was 2500 feet (750 meters). Actually, this gorge cuts across an even older filled gorge, the Martinez Gorge, but lack of space prevents its portrayal in Figure 9.

A younger gorge, the Princeton Gorge (Figure 9, PG), extends south and southwest from Chico, about 90 miles (144 kilometers) north of Sacramento, to beyond Capay, west of Sacramento. At Capay, the gorge is 2000 feet (610 meters) deep.

The youngest of the gorges, the Markley Gorge (Figure 9, MG), persisted into Oligocene time, as indicated by Oligocene marine sediments at the top of the gorge filling. The Markley Gorge has been traced from Marysville southward beneath Sacramento to Rio Vista in the Sacramento delta area. Its dimensions are comparable to the other gorges. It is possible that still other gorges lie hidden in the thick sedimentary fill of the

Great Valley. These gorges probably had tributaries entering from the mountains on either side, but the details are unknown.

Meanwhile, in the Sierran foothills, the streams at times eroded their valleys and at other times cluttered them with sediment. The changes in behavior were probably due to a variety of causes: fluctuations of sea level in the Great Valley, tectonic uplift and depression, volcanic activity, and climatic changes. Deep valleys were eroded in the foothills during the Eocene Epoch, and these were later widened considerably. The Eocene climate was humid-tropical or subtropical, as indicated by the presence in the foothill landscape of a thick reddish soil locally cemented to a resistant crust and similar to the laterite soils of present-day humid-tropical and subtropical environments.

The deep foothill valleys were floored with coarse gravels averaging about 50 feet (15 meters) thick but reaching a thickness of several hundred feet in the valley of the Yuba River. The upper layers of these gravels are white and consist primarily of quartz. They contain considerable quantities of fine gold released from the parent rock upstream by intensive weathering under warm, moist climatic conditions. These are the most important gold-bearing gravels in the Sierra. Locally, in Butte County and elsewhere, small diamonds have also been found in these gravels. The intensity of weathering is indicated not only by the destruction and removal of all minerals except the weather-resistant quartz, but also by the deep tropical soil mentioned earlier. Much of the gold was released from the Mother Lode, part of a belt of gold-rich quartz veins about one mile wide along a fault system extending more than 120 miles (190 kilometers) from Mariposa, west of Yosemite, to Forest Hill on the North Fork of the American River. California Highway 49 follows this gold-rich belt for much of its length. North of the Mother Lode belt, the source of the placer gold was the widespread system of veins in the rocks adjacent to the granite batholiths. This is true, for example, of the gold of the Grass Valley district in the Sierra east of Marysville. Gold-bearing gravels of this source are found almost to the present range divide. In the southern Sierra, the absence of gold veins is due

39

to deep erosion of the gold-bearing rocks, perhaps because they lacked the protection of a volcanic cover.

The early gold-bearing gravels of the northern Sierra are known as the prevolcanic gravels for reasons that will soon become obvious. They are the oldest gold-bearing gravels in the Sierra and are now largely concealed by later deposits. Later downcutting by the Sierran rivers has left these gravels stranded high on the present valley sides.

The vigor of the gold-gravel streams diminished as the streams approached the gentler slopes of the lower foothills and the border of the Great Valley. Only the more easily transported sands, silts, and clays were carried this far. These fine sediments were deposited along the shores of the interior sea or were swept into it to form large deltas. During low stands of the sea, the rivers trenched the deltas and probably generated turbidity currents on the submerged delta slopes. These turbidity currents may account for the erosion of the seaward portions of the deep submarine gorges mentioned earlier.

The Ione Formation, named after the town of that name southeast of Sacramento, consists of clays and silts, the downstream equivalents of the gold-bearing gravels in the Sierra. The Ione Formation is exposed intermittently along the east side of the Sacramento Valley, where it rests directly on the reddish soil mentioned earlier. The formation includes layers of commercial potter's clay and beds of coal. The copious vegetation represented by the coal beds supports the evidence provided by the reddish soil for a humid tropical or subtropical climate during the Eocene Epoch. These littoral and swamp conditions prevailed at other places around the borders of the Great Valley. Thus, Eocene deposits in the Mount Diablo area also contain beds of commercial clay and layers of coal. Coal mining on the east side of Mount Diablo centered around four towns abandoned about the turn of the century.

The gravel-laden Sierran streams emptying into the San Joaquin Valley also carried sands, silts, and clays into the interior sea. Because of the absence of a Mother Lode higher upslope, however, there were no gold-bearing gravels.

Across the Great Valley, the streams draining the Coast Ranges were small and the areas they drained were largely composed of relatively weak rocks incapable of providing resistant pebbles. As a result, gravels were only deposited locally along the west side of the valley, and then only in thin beds. Sands, however, are common.

It is uncertain how far the Coast Ranges extended seaward at this time, because the evidence west of the San Andreas Fault has long since moved north out of our area.

To the south, the Point Reyes Peninsula, the Santa Cruz Mountains, and the Gabilan and Santa Lucia ranges had not yet reached Middle California. One thing is clear: during their migration north, these ranges were at times partially submerged, for early Tertiary marine sediments are preserved within their limits. In the Santa Cruz Mountains, the marine sediments are more than a mile thick, suggesting a continuously subsiding basin. Some of these early Tertiary rocks may be seen along California Highway 1 at San Pedro Point on the coast south of San Francisco and in Point Lobos State Park south of Carmel. At San Pedro Point, the sediments are interbedded shales and sandstones with conglomerates only at the base; at Point Lobos, conglomerates predominate.

After deposition of the prevolcanic gold-bearing gravels in the Sierra, volcanic eruptions began on a grand scale in the higher parts of the range. The eruptions began in the Oligocene about 33 million years ago, and terminated in the Miocene about 16 million years ago. The activity, however, was intermittent; there were long pauses during which streams eroded the volcanic deposits and soils and vegetation accumulated.

The centers of volcanic activity were in the higher parts of the ancestral Sierra. Vast quantities of light-colored ash were exploded into the atmosphere, and considerable volumes of light-colored lava were extruded. The lava, equivalent to granite in composition, is known as *rhyolite*. Many of the violent eruptions spewed forth enormous clouds of incandescent ash which, because of their weight, rushed down the Sierran slopes in the form of fiery clouds known as *glowing avalanches*. Modern examples of glowing avalanches attest to their destructiveness. In 1902, a glowing avalanche rushed down the flanks of Mt. Pelée on the island of Martinique in the West Indies and overran the city of St. Pierre, killing all but two of the 28,000 inhabitants.

The rhyolitic lavas, ashes, and inevitable mudflows eventually reached a thickness of over 1000 feet (300 meters) in the higher parts of the Sierra, where they buried all but the highest hills as shown in the upper left of Figure 9. The ash, of course, was spread more widely by the wind and settled among the fluvial and marine deposits of the Great Valley. In the foothills, the gravel-floored valleys were overrun by glowing avalanches and partially filled with ash and lava. Between eruptions, and after cessation of volcanic activity, streams eroded the volcanic deposits and mingled the debris with the debris of other rock types beyond the limits of the volcanic terrain. Thus, much of the volcanic-rich alluvium of the foothills and Great Valley is equivalent in time to the rhyolitic volcanics in the higher Sierra.

Geologists know these volcanic rocks as the *Valley Springs Formation*. Note in Figure 9 that the volcanics are not found south of the Merced River drainage system. They may have extended a little farther, but if so, all traces have been removed by erosion.

On the right side of Figure 9, we see the prevolcanic Sierran drainage pattern forced into a trellislike pattern by the parallel ridges of metamorphic rock rising above the granite lowlands. In the absence of an accumulating volcanic cover, erosion continued without interruption here. Major valleys were opened out to widths of several miles, the *Broad Valley Stage* of Sierran erosion.

On the left side of Figure 9, the accumulating volcanics have largely buried the original drainage pattern. On completion of the volcanic plain, the rivers followed more direct courses down the western slope.

Meanwhile, the climate was slowly changing. It began to grow colder in late Eocene time and the chilling continued into the Oligocene. Glaciers probably accumulated in polar regions, thereby lowering sea level. The resulting widespread erosion may explain the present absence of marine Oligocene sediments in the Great Valley except in the deeper portions including the gorges described earlier.

Early Tertiary time closed with a widespread retreat of the sea from Middle California. At least part of the withdrawal may be related to the glaciation mentioned above, but part may be due to regional elevation of middle California. In any event, terrestrial sediments spread widely into the Great Valley following the shrinking inland sea. The deep gorges, however, were still under water, and those that had not yet been completely buried were still being filled.

MIOCENE TIME

21 to 5 million years B.P.

Figure 10 synthesizes the major events that took place in Middle California from the activation of the San Andreas Fault about 21 million years ago to the episode of mountain-making that terminated the Miocene Epoch, 5 million years ago. Remember that the San Andreas Fault System first originated in southern California when the East Pacific Rise, moving east between transform faults, descended under the edge of the continent and subduction ceased. The oceanic plate, however still moving to the north, tore off and carried with it the overlapping edge of the continent. This tear marked the beginning of the San Andreas Fault System.

During the long interval of time telescoped in Figure 10, the sea first invaded the Great Valley at its extreme southern end. The rest of the Great Valley was then above water and blanketed by terrestrial sediments washed in from the bordering mountains. As the sea spread northward, it submerged the terrestrial sediments and covered them with marine deposits. The expansion of the sea was not continuous; there were pulsational advances and withdrawals clearly indicated by the interlayering of marine and terrestrial sediments.

Toward the close of the Miocene, the interior sea reached almost to Sacramento, and an extension probed as far north as the lower end of Clear Lake (Figure 10, L). The sea in the Great Valley probably connected with the open ocean just south of the present site of San Francisco Bay. For convenience, we shall refer to this connection as the San Jose Strait.

Deciphering the evolution of the westernmost strip of the Coast Ranges during the remainder of the Tertiary period presents problems. The landscape we see today west of the San Andreas Fault was far to the south in late Oligocene time.

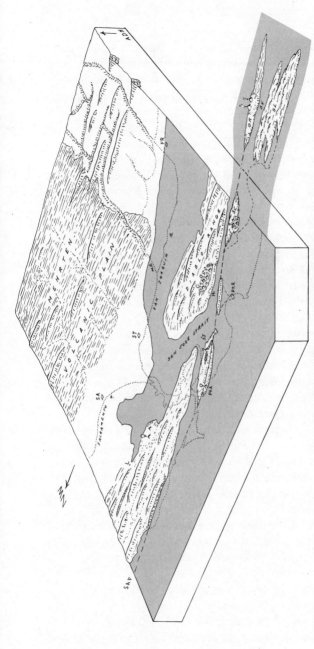

Figure 10. Early to late Miocene, 21 to 5 million years B.P. The Mehrten volcanics buried and extended beyond the earlier Valley Springs volcanic deposits. The inland sea, at its maximum, reached to about the latitude of Sacramento. At its maximum, the sea probably connected to the open ocean through San Jose Strait. Volcanism was widespread throughout the Coast Ranges, the largest area being east of Hollister (H). Volcanic activity erupted in the Gabilan Range (GR) at the present site of Pinnacles National Monument. West of the San Andreas Fault, the forward elements of the moving oceanic block had entered Middle California. (FR—Fresno; GR—Gabilan Range; H—Hollister; MR—Merced River; Pt.A—Point Arena; Pt.R—Point Reyes; SA—Sacramento; SAF—San Andreas Fault; SC—Santa Cruz Range; SL—Santa Lucia Range; ST—Stockton; SV—Salinas Valley.)

Much of the development of the exposed portions of the moving coastal block was accomplished during the long migration
north into our area. Among the present-day features that were
carried along with the moving block were the Santa Lucia
Range, the Gabilan Range, the Santa Cruz Range, the Point
Reyes Peninsula, Bodega Head, and the long coastal sliver
between the mouth of the Russian River and Point Arena.

We have no way of knowing what the exact configurations
of the exposed segments of the moving slab were during the
migration north. We are able, however, to determine whether
the segments were larger or smaller, based on the extent of
marine deposits within them. But other important evidence is
undoubtedly submerged offshore. We shall therefore portray
these moving landscape features in our sequential diagrams
much as they appear today. It bears repetition, however, that
these segments were both sliced and diced as they moved
north; sliced by faults parallel to the San Andreas Fault, and
diced by many cross-faults in the intervening segments.

Starting with Figure 10, we shall exercise a little poetic
license and show the coastal landscape features of Middle
California in the various positions they may have occupied
during their migration north along the fault. In each figure,
therefore, those portions of the Middle California coastal strip
that had not yet entered the area are shown in suspended fashion off the lower right side of each diagram. In each succeeding diagram, these features as well as those already within the
confines of the diagrams are shown in progressively more
northerly positions until they arrive at their present sites.

As for the speculative strip of coastal landscape depicted in
Figure 9 on the seaward side of the San Andreas Fault, it too
shifted north as the fault became active. These features have
long since moved out of our area and were probably mashed
under the margin of the continent in areas of still active subduction to the north.

Our sequential diagrams are based on the precarious assump-
tion that the oceanic slab west of the modern San Andreas Fault
System has moved northwest 190 miles (300 kilometers) in 21
million years, or almost 10 miles (16 kilometers) per million

years. We know, however, that the movement was not restricted to a single fault, but progressed northwest transferring from one fault to another. Because the complexities are largely unknown, and for the sake of simplicity in our story, we will consider the San Andreas Fault as the only locus of movement rather than the many faults that make up the San Andreas Fault System.

For the above reasons, the position of the moving landscape features shown in Figure 10 is speculative. For example, Point Arena, instead of being opposite San Francisco, may have been some distance to the north or south. But that it was considerably south of its present location is indisputable. Similarly for Point Reyes, the Santa Cruz Mountains, the Gabilan Range, and the Santa Lucia Range.

We have already commented on the strong probability that the configuration of the Santa Lucia, Santa Cruz, and other ranges of the northward-moving oceanic segment changed considerably during migration. After all, these ranges were themselves subjected at various times during their migration to tectonic forces which crumpled and shattered them and which elevated or depressed them in whole or part. And fluctuations in sea level contributed to their changing dimensions and shapes. Most of these landscape features have been out of the sea for a long time, however, for they include large areas of sediment that were laid down on dry land.

It is equally clear that Monterey Bay is of ancient vintage. It began to develop in the late Oligocene Epoch; there was already a basin here when the Miocene seas invaded the area, as indicated by the presence of marine Miocene sediments under the entire basin and lapping against the flanks of the surrounding mountains. Furthermore, a deep submarine canyon in Monterey Bay has an ancient buried continuation extending inland to the head of the basin. This buried canyon is filled with and covered by marine Miocene sediments. Actually, Monterey submarine canyon was initiated while Monterey Bay was far to the south, for the bay, too, is part of the coastal block moving north along the San Andreas Fault. There is some evidence to suggest that the lower portion of Monterey Canyon has been

47

sliced and offset several times by San Andreas-type faults under Monterey Bay.

The Santa Lucia Range was smaller than now during its migration north and was probably an island. This is suggested by the presence of marine Miocene sediments, known as the Monterey Formation, around much of the borders of the range. They are widely exposed on both sides of Carmel Valley at the north end of the range. Near Carmel Mission along Highway 1, road cuts reveal dipping beds of the Monterey Formation resting on granite. The beds were, of course, originally deposited horizontally in the Miocene sea; the tilting was the result of later upheavals. But the presence of these beds indicates that at that distant date the northern part of the range lay beneath the sea. Most of the Monterey deposits have been eroded from the steep western flank of the range, but a few remnants are preserved in downfaulted slivers. Marine beds are widely exposed in the south, near San Simeon and Hearst Castle, and for long distances along the east side of the range. All these peripheral deposits support the view that the range was an island, or perhaps a group of islands. The latter possibility stems from the occurrence of large areas of marine Miocene beds within the interior of the range.

Eastward across the seaway that occupied the site of the present Salinas Valley rose another long island where the Gabilan Range now stands. On this isolated island a succession of interesting events took place about 23 million years ago. There emerged from the depths, by way of great cracks in the Earth's crust, large volumes of porridgelike lava that spread over the surface, building up a long dome several thousand feet high. On the crest of this dome a string of volcanoes came into existence. At times these volcanoes erupted violently, scattering enormous quantities of broken rock over the area; at other times they erupted quietly, emitting great floods of lava. From this beginning, the scenic Pinnacles National Monument was eventually to develop.

Farther to the northwest along the migrating coast was another island, an ancestral, smaller version of the Santa Cruz Range. Its eastern flank coincided with the San Andreas Fault valley, then a narrow seaway. The northern part of this island was eventually to become part of the San Francisco Peninsula. Of special interest was the presence of active volcanoes in the La Honda area about 10 miles (16 kilometers) southwest of Stanford University. Lavas from these volcanoes flowed westward into the sea and may now be seen above water covered by later seafloor sediments. Ash and cinders within the sediments indicate that the volcanoes at times erupted explosively. These are the oldest post-Jurassic volcanic rocks in the Coast Ranges.

Much of the molten material that rose from the depths never broke through to the surface, but instead sandwiched itself in among the layers of sediment that were accumulating on the sea floor. Thanks to later uplift and erosion of the Santa Cruz Mountains, we can see these ancient rocks in many road cuts and stream banks in the higher parts of the range. The best exposures are at Langley and Mindego Hills, about 12 miles (19 kilometers) southeast of the town of Half Moon Bay.

Some time during this volcanic episode, a volcano was built up east of the San Andreas Fault about 2 miles (3 kilometers) south of Stanford University. During violent eruptions, cinders from this volcano rained down in the nearby sea. At one time during the life of this volcano, a stream of basaltic lava flowed northwest for about 3 miles (5 kilometers). The lava was about 400 feet (120 meters) thick near the volcano but only 20 feet (6 meters) thick where it is penetrated by a well at its northern end. The lava is exposed in the quarries along the old Page Mill Road near Stanford.

The Monterey sediments that were deposited in the sea around the flanks of the Santa Cruz Range are now above water in many places. Near Santa Cruz itself they have been battered by the sea and carved into narrow promontories—thin partitions between adjacent coves. In places, the promontories have been worn through to create arches and natural bridges. This is the site of Natural Bridges Beach State Park (Photo 10).

Southeast of the present site of the San Francisco Bay lowland lay a broad sprawling upland area. This was an early ancestor of the Diablo Range, but it extended only from the lower end of the present bay to about the latitude of the Pinnacles. The entire north end of the range was not then in existence; the waves of the Pacific washed across the area where now lie the Contra Costa and Diablo hills.

Volcanoes were active in scattered areas of the ancestral Diablo Range. The largest volcanic area was east of Hollister where Cathedral Peak and many other peaks in an area of about 180 square miles (470 square kilometers) are carved out of a complex of ancient lava flows and intrusive rocks.

The Point Reyes Peninsula, which by late Miocene time had reached the present site of Monterey Bay, was largely under water, and sediments of the Monterey Formation, similar to those that were deposited elsewhere in this ancient sea, were laid down. The sediments consist largely of shale, a rock formed from the compaction of mud.

North of San Francisco Bay, the late Miocene seas extend inland as far as the Sonoma Range; remnants of the marine sediments are found along Carneros Creek, 5 miles (8 kilometers) west of Petaluma, and oil seeps in the Sonoma Mountains are believed to have their source in deeply buried Monterey sediments.

The configuration of the shoreline of the ancient Miocene sea north of San Francisco Bay is not precisely known. Marine sediments are absent north of the inundated area shown in Figure 10, but they could have been eroded away after deposition.

There were centers of volcanic activity north of San Francisco Bay as well as to the south. As a matter of fact, volcanic activity was more widespread during the Miocene than at any time since. Scattered volcanoes north of the bay erupted intermittently and distributed volcanic ash over the adjacent seas. Ash from these northern volcanoes is found in the marine sediments as far south as Mt. Diablo. And at Pt. Arena, which at this time was about opposite Golden Gate, more lava poured out. Whether this lava came from a long-vanished volcano or from a crack in the Earth's crust is unknown. The volcano shown there in Figure 10 is not completely authenticated.

The highland areas of Miocene time were, of course, subject to erosion, and the sediment-laden streams carried the debris down into the sea. By late Miocene time, the lands were worn nearly to sea level. Even the volcanoes that dotted the region were worn away. This episode of erosion was widespread throughout the Coast Ranges and may have been worldwide.

Meanwhile, in the Sierra, the volcanic activity of the Valley Springs episode continued unabated into the Miocene. Mudflows and lava flows spread far down into the foothills, where the deposits were eroded by streams and carried down into the Great Valley. At the same time, volcanic ash drifted far afield, settling in the accumulating sediments of the Great Valley and even as far as the Coast Ranges.

Valley Springs volcanism was intermittent; during periods of quiescence, deep valleys were eroded in the volcanic deposits. In these valleys, other gold-bearing gravels accumulated, only to be buried by later volcanic deposits. These gold-bearing gravels were not nearly as rich as the Eocene prevolcanic gravels.

The Valley Springs volcanic episode terminated about 16 million years ago in the early Miocene. A long period of quiet ensued, during which the Sierra was deeply eroded with a relief of several thousand feet. In the granitic Sierra, south of the volcanic terrain, the rivers cut deep, steepsided valleys below the floors of the earlier broad valleys. This episode of erosion is known as the *Mountain Valley stage*.

And now a new episode of volcanism began in the northern Sierra. It differed from the Valley Springs episode in that the lava and ash emitted were darker in color and different in composition from the Valley Springs volcanics. For convenience we will refer to this episode as the Mehrten volcanic episode, although this name has been applied in the foothills to only part of these deposits.

The Mehrten volcanic deposits were piled up more than a mile thick in the higher parts of the Sierra but thin to only a few hundred feet in the foothills. Their distal remnants probably lie buried under the later deposits of the Great Valley. Within the Sierra, much of the material consists of mudflows resulting from saturation of ash deposits on the volcanic slopes.

The Mehrten volcanics completely buried the earlier foothill topography north of the latitude of Merced except for the highest ridges and a few isolated peaks (see Figure 10). On this volcanic cover a new generation of Sierran streams came into existence, flowing directly westward down the volcanic slope except where deflected by the foothill ridges. The direct westward flow was aided by early tilting of the Sierra Nevada, although the major tilting came later. South of the Mehrten volcanic plain, the original right-angle trellis pattern of the Sierran streams persisted.

After most of the Mehrten volcanics had accumulated, a long period of quiescence ensued. Deep valleys were eroded through both the Mehrten and Valley Springs volcanics and on into the basement rocks below. These valleys were also floored with gold-bearing gravels, but these gravels, too, were not as rich as the Eocene prevolcanic gravels.

Then, about 9 million years ago, streams of dark lava poured down some of these deep valleys from centers of eruption along the crest of the ancestral range. They covered the floors of the valleys, burying and protecting the gravels from further erosion. This was before the major faulting that raised the Sierras to their present great height.

The net result of the accumulation of the Valley Springs and Mehrten volcanic deposits was to convert the landscape of the northern Sierra to a smooth sloping plain interrupted only by the deep valleys of the major rivers with their lava-covered floors. During the long period of weathering and erosion that followed, the weaker volcanic deposits of the interstream divides, consisting largely of ash and mudflow deposits, were eroded more rapidly than the lava-covered valley floors. In time, the lava-protected valley floors were left standing high above the more deeply eroded surroundings; a complete reversal of topography. As a result, some of the old, winding valley floors now appear as high, sinuous, flat-topped ridges wandering through the countryside. Tuolumne Table Mountain (Photo 7) represents the high abandoned floor of ancient Tuolumne River. The course of Tuolumne Table Mountain in the foothills east of Knights Ferry is shown in Figure 1.

The granitic southern Sierra was also being eroded at this time. The foothill ridges, in the absence of surrounding volcanics, remained high and prominent. And as erosion progressed in the granitic terrain, the distribution of fractures began to exert a strong influence on the developing topography. In many places, the fractures were concentrated in narrow zones and so weakened the rock as to lead to accelerated valley erosion. And because fractures commonly occur in cross-cutting sets, erosion commonly left a central core of massive, unfractured granite surrounded by deep valleys. These cores were eventually to develop into the picturesque domes of the High Sierra. Although they probably made their first appearance at least as early as late Miocene time, they may not have become conspicuous features until much later. Hence, the details of their origin will be considered in the discussion of Pliocene events.

Meanwhile, in the Coast Ranges, episodes of deposition alternated with episodes of crustal folding and erosion. For example, the thick Miocene sediments of the Santa Cruz Mountains did not accumulate uninterruptedly; instead, in many places, earlier sequences were tilted up and eroded before the next sequence of beds was laid down. These alternating episodes, however, did not occur simultaneously throughout the Coast Ranges. Deformation was spotty in space and time. Some areas were uplifted while neighboring areas were depressed, and the roles were often reversed at later times. There were times, however, when deformation seems to have been regional in scope.

In Figure 10, the Coast Ranges are shown as they may have looked after the long period of erosion that preceded deposition of the latest Miocene sediments and prior to the strong crumpling of these sediments at the close of the Miocene. A few upheaved welts are shown to suggest the nature of the widespread deformation that followed.

As for the role of the San Andreas Fault during this 25-million-year interval, the first displacements, in southern California, probably took place along a member of the fault system at the edge of the continental shelf. As additional, more northerly portions of the East Pacific Rise reached and plunged under the continent, terminating subduction, the tearing of the coastal strip spread northward, probably along faults progressively farther shoreward. About 15 million years ago, subduction ceased at about the present site of the Santa Lucia Range, and the coastal strip was being displaced along faults approaching closer and closer to the San Andreas Fault. By the end of the Miocene, 5 million years ago, subduction ceased at about the latitude of San Francisco, and the coastal strip was being displaced along the San Andreas and neighboring faults. The site of cessation of subduction and initiation of lateral displacement has continued to shift northward during the succeeding 5 million years to the present. Subduction has now ceased as far north as Cape Mendocino, but is still active beyond.

Late Miocene Mountain–Making

For some obscure reason, the subterranean forces that had been intermittently active throughout much of the Tertiary Period began to stir again toward the close of the Miocene and caused relatively rapid changes in the landscape. Of course, from our point of view, the changes were extremely slow and probably passed unnoticed by the creatures that lived at that time.

The pressures exerted by these forces created great welts here and there in the Coast Range landscape. In time these grew into giant folds, like wrinkles in a loose carpet. A few of these welts were indicated in Figure 10. Earthquakes were common, the result of the shifting about of blocks of the crust. Most of the major faults of the region, such as the San Andreas Fault, were already in existence. New volcanoes came into being because faults offer easy upward passage for molten material from the depths. The earthquakes and volcanic activity were the growing pains of the rising mountains.

This was a time of upheaval, of crustal unrest. The sediments that had been deposited previously in horizontal layers on the Miocene sea floor, and those that had been laid down in interior basins, were crumpled into folds several thousand feet high and miles across. Locally the folded rocks were heaved up bodily along faults. And these changes in the landscape took place to the accompaniment of volcanic eruptions.

During this period of crustal deformation, the Miocene sediments that had been deposited in the seas surrounding the Santa Lucia Range were lifted above sea level, thereby increasing the dimensions of the range. The Santa Lucia Range had not yet reached our area of interest; its approximate position is shown in Figure 10.

The Gabilan Range across the Salinas Valley had been eroded down to its granitic roots prior to the volcanic eruptions described earlier. The eruptions continued into the late Miocene. Like the Santa Lucia Range, the Gabilan Range was elevated and enlarged at the close of the Miocene. Part of the uplift took place along faults, including the San Andreas Fault along the east side of the range. Much of the area of Gabilan Mesa to the south may have been above sea level throughout this period, because no exposures of marine rocks are revealed.

An especially interesting episode in the history of the Pinnacles area now took place. Great cracks developed across the volcanic dome that had been built here earlier, and a large segment of the dome, 2 or 3 miles (3 to 5 kilometers) wide and more than 6 miles (10 kilometers) long, subsided below the surface. The location of these parallel faults is indicated in Figure 12. A block of the earth's crust that drops down between faults (Figure 11) is known as a graben (also the plural form). The valley of the Dead Sea and the Rhine Valley in its most picturesque portion are modern examples on a grand scale. Had the Pinnacles graben not formed here about 10 million years ago, there would be no Pinnacles National Monument today. Because of this subsidence, the volcanic

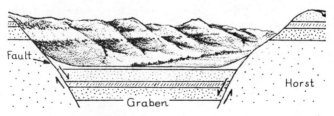

Figure 11. Graben and horst.

rocks out of which the spirelike pinnacles were later carved were dropped deeply enough to escape complete removal in the episodes of erosion that followed.

The Monterey Bay lowland, as previously noted, was already in existence, but the confining mountains had not reached their present dimensions. The elevation of the Santa Lucia Range to the south and the Santa Cruz Mountains to the north brought the basin into prominence. At the same time, the floor of Monterey Bay continued to subside, carrying the Miocene marine sediments to progressively greater depths.

The distribution of the marine Miocene sediments north of ancient Monterey Bay and east of the San Andreas Fault indicates that a long narrow seaway existed on the east side of the Santa Cruz Mountains, still not yet completely into our picture. A similar narrow inlet submerged the present site of Hollister, behind the migrating Point Reyes Peninsula. Still another narrow seaway extended north along the west side of the present San Francisco Bay lowland. Point Arena at this time may have been just outside the present Golden Gate.

The Diablo Range, like the other ranges, grew in size by folding and uplift not only of part of the surrounding sea floor but of the margin of the vast alluvial plain that extended into the Great Valley. The steep western margin of the Diablo Range was the result of uplift along another great fracture known as the Hayward Fault.

North of San Jose Strait, the crustal block out of which the Marin Mountains and the Mendocino Range were subsequently formed was uplifted bodily above the sea and parts were crumpled into broad folds and locally faulted. The upheaval of the western part of this block took place along the San Andreas Fault. Other faults cracked the region between the Marin-Mendocino block and the Sacramento Valley.

The Great Valley, too, was elevated sufficiently high to drive the seas out of the interior. And the Sierra was tilted up further, invigorating the west-flowing streams.

1. Sutter Buttes, also known as Marysville Buttes, about 40 miles (65 kilometers) north of Sacramento. View west. A Pliocene dome about 10 miles (16 kilometers) across, surmounted by scattered Pleistocene volcanic cones. The summits rise 2100 feet (650 meters) above the valley floor. (Photo by Pacific Aerial Surveys, Oakland.)

2. View east over San Francisco Bay region. San Francisco in right foreground with Golden Gate Bridge joining it to Marin Peninsula on left. Across the bay are Berkeley and Oakland, with Mount Diablo in left background. The main part of the Diablo Range appears in right background. In the far distance, beyond the Great Valley, are the high peaks of the Sierra Nevada. (Photo by Clyde Sunderland, Oakland.)

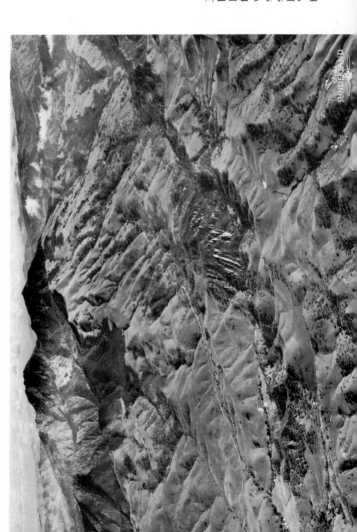

3. Mount Diablo in the Coast Ranges east of San Francisco Bay. The high peaks are eroded from a core of dark igneous rock which punched up through the sedimentary cover. Note the steeply dipping sedimentary layers near the center of the view. View toward southeast. (Photo by Clyde Sunderland, Oakland.)

4. (right) Mount St. Helena, at the head of Napa Valley, north of San Francisco Bay. The peak consists of rhyolitic volcanic rocks of Pliocene age. (Photo by Clyde Sunderland, Oakland.)

5. (below) Calaveras Formation exposed at Geologic Exhibit along Yosemite Highway between Briceburg and Chinquapin. (Photo by Arthur D. Howard.)

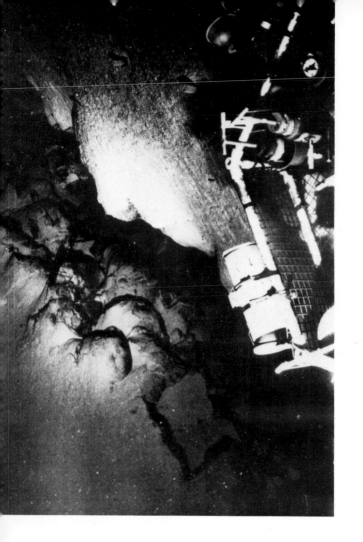

6. Open fracture in depths of Mid-Atlantic Ridge 350 miles (550 kilometers) west of the Azores. Depth, 9000 feet (2750 meters). Note pillowlike basalts exposed in wall of fracture. Project FAMOUS (French American Oceanographic Undersea Study). (Photo courtesy of Tjeerd H. van Andel.)

7. Table Moutain in Sierran foothills capped by a lava flow which flowed down the gravel-covered floor of an ancient Sierran valley, the precursor of the modern Stanislaus River Valley. The landscape on either side, being composed of weaker deposits, has been lowered by erosion to leave the former valley floor standing as a sinuous flat-topped ridge. View northeast from between James-town and Melones, about 3 miles (5 kilometers) west of Sonora. (Photo by John S. Shelton.)

8. Mariposa Slates along Yosemite Highway about 17 miles (25 kilometers) west of Mariposa. (Photo by Arthur D. Howard.)

9. Whalers Cove, Point Lobos State Park. View east. Marine terrace in middle ground. Granite on left; Paleocene sediments on right. (Photo by Arthur D. Howard.)

10. Sea arches in promontory of broad terrace eroded in Miocene Monterey Formation. Full width of terrace visible in background. Natural Bridges Beach State Park, Santa Cruz. (Photo by Arthur D. Howard.)

11. San Andreas Lake in foreground along San Andreas fault zone. San Bruno Mountain in distance, an eroded fault block tipped up during the mid-Pleistocene mountain-making. The broad intervening lowland is underlain by the Plio-Pleistocene Merced sediments. (Photo by Arthur D. Howard.)

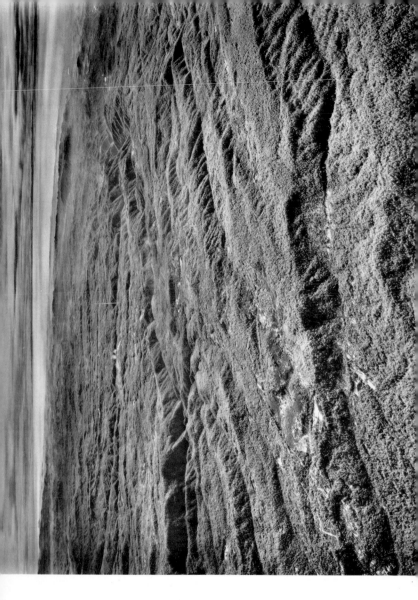

12. Backslope of the northern Sierra Nevada. View toward south. This sloping plain, with its many broad flats, is the surface of the Mehrten volcanic plain. In the distance are the foothill ridges of Jurassic serpentine. Canyon of Cosumnes River in foreground; canyon of Mokelumne River in middle distance. (U.S. Geological Survey photo GS-OAH-6-68.)

13. Farallon Islands, the wave-eroded crest of a granitic fault block mountain. The hills rise from a marine terrace.

14. View in Pinnacles National Monument at south end of Gabilan Range. Eroded from Miocene volcanic rocks.

15. View northwest across Clear Lake. Mt. Konocti (upper left), a late Pleistocene compound volcano, rises 3000 feet (900 meters) above the lake. Borax Lake, the first commercial source of borax in the United States, is in right center. Sulphur Bank Point, beyond Borax Lake, has produced considerable mercury and sulphur. The plain in the foreground is underlain by the Plio-Pleistocene Cache Formation. Volcanic activity in this area has continued until recent times, and hot springs and fumaroles are widespread. See Figure 18 for additional locality names. (Photo courtesy of *San Francisco Examiner*.)

16. Geyser steam area about 15 miles (24 kilometers) south of Clear Lake. Now a major source of geothermal energy. (Photo by Arthur D. Howard.)

PLIOCENE TIME

EARLY TO LATE PLIOCENE

5 to 2 million years B.P.

The Coast Range folds created in the late Miocene orogeny, representative examples of which are shown in Figure 10, were reduced by erosion to a rolling plain surmounted by many residual ridges (Figure 13). Irregular warping of this landscape permitted the sea to embay large parts of the coastal region, and differential warping of the floor of the Great Valley enabled the sea to invade the southern part of the valley.

The sea that invaded the Great Valley encroached from the south and spread gradually northward until it reached the latitude of Fresno, just inside our area. North of Fresno, streams from the Sierra and smaller streams from the Coast Ranges spread alluvial sediments over the floor of the valley.

The streams of the Sierra Nevada, rejuvenated by the late Miocene tilting, began to trench the Mehrten volcanic plain in the north and accelerated their valley deepening in the granitic and metamorphic terrain to the south. In the higher parts of the Sierra, the accelerated erosion resulted in the excavation of steep, V-shaped canyons below the floors of the more open Miocene mountain valleys. The mountain valleys, it will be recalled, had earlier been excavated below the floors of Oligocene broad valleys (Figure 12). The renewed erosion produced huge amounts of sediment which were carried down into the Great Valley and spread out as alluvial fans. As the fans increased in size, they merged to form an extensive alluvial plain reaching far out into the valley. For convenience, we will refer to this plain as the Laguna Alluvial Plain (Figure 13), after the Laguna Formation in the northern San Joaquin Valley. Actually, the sediments of this alluvial plain are known by different names in different parts of the valley.

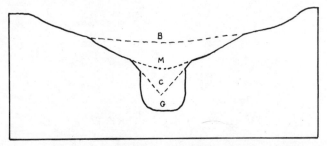

Figure 12. Valley-in-valley profile, Sierra Nevada. During the Oligocene epoch when the Sierra stood low, the streams opened out wide valleys, the Broad Valley Stage (B). During the Miocene epoch, the rivers entrenched themselves and carved narrower, more steeply sloping valleys, the Mountain Valley Stage (M). In the Pliocene epoch another episode of rejuvenation resulted in another pulse of downcutting to give steep, V-shaped canyons, the Mountain Canyon Stage (C). Finally, in the higher Sierra, glaciers converted the V-shaped canyons to U-shaped glacial troughs (G).

As one would expect, the streams draining the Mehrten Volcanic Plain carried considerable volcanic debris into the Great Valley, whereas the streams draining the southern Sierra contributed granitic and metamorphic debris. Many of the modern stream courses in the northern and southern Sierra were initiated or modified during this period of time. Some of the northern rivers eroded their valleys deeply enough to penetrate the old buried placer gold deposits and reconcentrate the gold along their new channels. This was the origin of the placer gold at Sutter's Mill on the South Fork of the American River that led to the California gold rush of 1849.

Meanwhile, west of the foothills, about 40 miles (65 kilometers) north of Sacramento, a strange event was taking place. A low circular dome began to rise above the floor of the Great Valley (Figure 13). The mounding was caused by the slow intrusion, below the surface, of a local body of magma. The dome continued to grow until, as we shall see later, it experienced violent eruptive activity creating the present-day group of mountains known as Sutter or Marysville Buttes.

Farther west, across the Great Valley, streams from the Coast Ranges were also depositing an apron of debris along the valley margin, but of much smaller dimensions than the

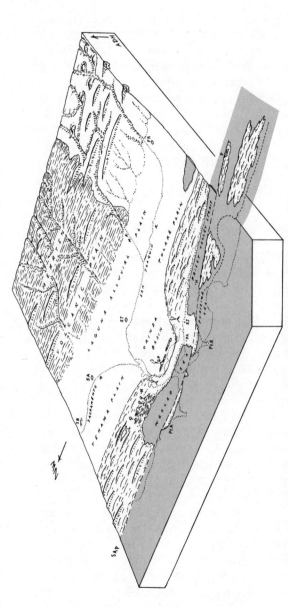

Figure 13. Early to late Pliocene landscape, 5 to 2 million years B. P. Erosion of the Sierra Nevada provided large quantities of sediment which were spread out in the Great Valley as the Laguna Alluvial Plain. South of the Mehrten Volcanic Plain, many granitic domes now made their appearance. The inland sea barely extended into Middle California at this time. A ridge now separated the Great Valley from the open sea. Sediment from the Coast Ranges formed the Tehama and Tulare plains and covered the floor of the Orinda Basin. The main center of volcanic activity had now shifted north of the San Francisco Bay region to the site of the Sonoma Range. The basin near Lower Lake (L) became the site of deposition of the Cache Creek beds. The Merced and Purisima seas indented the coast, and the latter connected with the inland sea by way of Priest Valley Strait. The Pinnacles volcanic rocks had been dropped down between parallel faults and preserved from erosion. (FR—Fresno; GP—Grizzly Peak; H—Hollister; L—Lower Lake; MR—Merced River; N—Napa; PA—Palo Alto; P—Pinnacles; Pt. A.—Point Arena; Pt. R.—Point Reyes.)

Laguna Alluvial Plain at the foot of the Sierra. In the Sacramento Valley portion of the Great Valley, the sediments have been named the Tehama Formation, and we shall refer to the plain in this area as the Tehama Plain (Figure 13). A layer of volcanic tuff near the base of the Tehama Formation was laid down about 3,300,000 years ago. South of the Bay region, an equivalent plain, the Tulare Plain, was being formed on the west side of the San Joaquin Valley.

It was in the vicinity of San Francisco Bay, however, that a relatively unique situation developed. Until late Miocene time, there appears to have been a gap in the Coast Ranges in this vicinity through which the Pacific Ocean had direct access to the Great Valley. During the late Miocene uplifts, however, this gap was apparently sealed off by a narrow ridge along the west side of the San Francisco Bay lowland, blocking the sea from the interior. Thus, on the east side of the ridge, terrestrial sediments were accumulating of the same age as the marine sediments in the sea on the ocean side. We shall refer to this basin of terrestrial deposition as the Orinda Basin, after the town of that name. The deposits themselves are known as the Orinda Formation. There were no Berkeley Hills or Mount Diablo at that time. The Orinda sediments entombed the remains of land plants and animals, including the bones of an ancestral three-toed horse. In places, plants grew so profusely that they eventually formed beds of coal. The sediments also include layers of volcanic ash, indicating volcanic activity in the vicinity. Inasmuch as the thickness of the ash diminishes to the south, the principal source must have been north of the bay.

Another interesting basin that developed during the Pliocene Epoch lay just east of present Clear Lake. This basin was a graben, dropped down between parallel northwest-trending faults. Within this basin there accumulated a thickness of more than a mile of sediment washed in from the surrounding areas. These sediments, named the Cache Creek Formation after the many exposures along the creek of that name, consist largely of stream-laid silts, sands, gravels, and, near the top, lake beds. Depositional features within the sediments indicate that the basin drained out to the northwest. The enormous thickness of alluvial sediments in the Cache Creek Basin proves that the basin was subsiding slowly and continuously. If the basin had dropped down all at once, or in a few precipitous steps, lake beds would have dominated, but they do not.

Toward the end of the episode of deposition of the Cache Creek deposits, several layers of volcanic tuff were laid down. These were the first symptoms of volcanic activity in the Clear Lake region. The major area of Pliocene volcanic activity, however, the Sonoma volcanic field, was just north of San Francisco Bay and east of the Petaluma–Santa Rosa lowland. At times, enormous volumes of lava emerged from volcanoes or deep fractures and inundated the countryside; at other times, large quantities of ash, cinders, and rock fragments were exploded over the landscape, contributing to its burial. The volcanic activity was not continuous, however. Sometimes the volcanoes lay dormant for long periods of time, and soils formed on the lavas and ash beds. Forests grew in these soils, only to be buried under subsequent floods of lava and ash. The trees of the Petrified Forest between Santa Rosa and Calistoga grew in one of these quiet periods. But during the eruption of a nearby volcano, heavy rains from moisture-laden clouds saturated the loose ash on the slops of the volcano and created rivers of mud. One of these flowed down into an adjacent lowland at the site of the present Petrified Forest. The fossil trees of the Petrified Forest, including huge redwoods and firs, have all been toppled to the southwest, proving that the flow of mud came from the northeast.

Much lava flowed down into the neighboring sea, and large amounts of ash settled into its depths. These were buried under the sediments collecting on the sea floor.

The end result of this prolonged period of volcanic activity was the accumulation of several thousand feet of volcanic rocks and mudflows covering an area of more than 350 square miles (900 square kilometers). Toward the close of this episode of volcanic activity, about 3 million years ago, hot ash flows spread over the countryside. One of these caps Mt. St. Helena.

There was also volcanic activity east of San Francisco Bay. Where the Berkeley Hills now stand, lava came from the depths on several occasions, probably from great fractures but possibly also from volcanoes. Some of the lava is light in color, as that at Northbrae on the west slope of the Berkeley Hills north of Berkeley. Lava was also extruded a short distance south of Berkeley and flowed southeastward for 21 miles (34 kilometers). This lava, exposed in the hills east of Alameda where it covers several square miles, is light bluish-green where freshly exposed, but yellowish or brownish where weathered.

Other lavas that came out at this time were almost black, like those that cap Bald and Grizzly peaks in the hills behind Berkeley.

There were apparently one or more volcanoes in the San Jose area, for the Pliocene sediments east of San Jose include volcanic debris; but the site of these volcanoes is unknown.

Note that the main area of Pliocene volcanic activity, the Sonoma volcanic field, is well north of the main Miocene field east of Hollister at about the latitude of Monterey Bay. This northward migration of the main centers of volcanic activity will continue into the Pleistocene Epoch.

Along the Pacific coast, a broad embayment extended inland over Marin County and the northern part of the San Francisco Bay region. The embayment extended north to the Mendocino Range, whose southern foothills lay north of the mouth of the Russian River. To the east, the embayment reached as far as the Sonoma Range. We shall refer to this large embayment as the Merced Sea. The marine deposits laid down in this sea have been named the Merced Formation.

South of San Francisco Bay, another Pliocene embayment flooded the present site of the southern Santa Cruz Mountains. We shall refer to this embayment as the Purisima Sea, after the widespread marine deposits that were spread over its floor. The Purisima Sea extended inland to the vicinity of San Jose, and a long narrow strait, the Priest Valley Strait, connected it with the interior sea occupying the San Joaquin Valley (Figure 13).

Meanwhile, the crustal segment west of the San Andreas Fault was continuing its march northward. By the close of the episode shown in Figure 13, Point Reyes was probably at about the present site of the Santa Cruz Mountains, and Point Arena had reached the present site of Point Reyes. The north end of the Santa Cruz Mountains lay just east of the present site of Monterey Bay, and the Santa Lucia and Gabilan ranges were just entering the southern boundary of our area.

The Point Arena sliver may have been above water at this time, because no marine Pliocene sediments have been discovered there. Of course, it is always possible that they might have been removed later by erosion.

Picturesque Inverness Ridge, which forms the eastern base of the triangular Point Reyes Peninsula, may also have been above water, but the greater part of the peninsula was submerged, as indicated by marine Pliocene sediments on top of the older Monterey Formation. The sediments consist of sandstones, siltstones, and shales. Their white color and relatively horizontal attitude near Drake's Bay so reminded Sir Francis Drake of the White Cliffs of Dover in 1579 that he named the region Nova (New) Albion, after the ancient name of Britain. These beds rest on the steeply dipping and eroded Monterey beds. The steep dips resulted from the late Miocene crustal disturbance, and the erosion surface that bevels across them was developed in the erosional interval that followed. The downwarping of this erosion surface enabled the lower Pliocene sea to invade the peninsula.

In the Santa Cruz Mountains, several cycles of uplift, erosion, depression, and inundation took place during the Pliocene. Such events were not contemporaneous over the Coast Ranges as a whole. It is clear that some areas rose while others remained static or subsided. This chaotic jostling was characteristic of the Coast Ranges. The jostling locally increased the deformation of the Monterey Formation, which had been tilted and folded in the late Miocene orogeny.

Volcanism continued sporadically during the Pliocene. In the Sierras, in the upper San Joaquin watershed east of the boundary of Figure 13, basaltic lava poured out on the floors of broad basins and locally spilled over the sides of the inner gorge of the San Joaquin River. These events took place about 3.5 million years ago. The cutting of the inner gorge obviously

63

began before these flows appeared. The deep erosion was apparently initiated by the tilting of the Sierra that began in earnest after the close of the Mehrten volcanic episode. The tilting involved little or no crumpling or folding, as evidenced by the smooth, flat, undeformed divides between major valleys in the northern Sierra.

We noted earlier that the crests of many of the picturesque granite domes of the Sierra first began to appear in the topography at least as far back as Miocene time. It was the major tilting of the Sierra, which began about 9 million years ago and which has continued to the present time, that brought these features into prominence. But why the domal forms? To understand this, we must bear in mind that the batholiths which form the Sierran landscape south of the Mehrten Volcanic Plain originally congealed miles below the Earth's surface. They were therefore under enormous pressure, the weight of the huge mass of overlying rock. As erosion gradually removed the overlying rocks, the pressure on the granite below diminished. The granite slowly expanded upward, first in its upper portions, then progressively lower. Since granite is relatively rigid, the expansion resulted in the shrugging off of plates of rock from a few inches to tens of feet thick. Now, we earlier mentioned that the Sierran granite is crisscrossed by sets of closely spaced fractures which encouraged the erosion of valleys along them. Between the checkerboard of fracture-controlled valleys, resistant masses of granite remained. The removal of rock from the valleys also reduced the lateral pressures which confined the cores. The cores therefore expanded toward the valleys as well as upward, resulting eventually in the splitting off of curved shells. This process is known as *exfoliation*. Much of this exfoliation occurred while the granites were still below the surface. Most of the domes of the Sierra are probably of this type, brought into being as valleys were eroded below the upland surface. Some, however, may be much older and may have been domal prominences rising above the upland surface below which most of the domes developed.

LATE PLIOCENE MOUNTAIN-MAKING

2 million years B.P.

The comparative quiet of the Pliocene Epoch, broken only by scattered volcanic eruptions and by local folding and fracturing, came to an end about 2 million years ago. The forces that had earlier acted sporadically, raising a welt here, depressing a basin there, were now more continuous and caused widespread changes in the landscape (Figure 14). The entire region was elevated bodily. In addition, much of the area of the Coast Ranges was thrown into folds and locally faulted. Areas between the folds, as well as isolated basins, became sites of deposition. One of these depressed basins, the Livermore basin, lay south of Mount Diablo. Another developed just west of the earlier Cache Creek Basin, where the Clear Lake volcanic field now lies. At least part of the San Francisco Bay lowland originated at this time. Many other smaller basins appeared throughout the Coast Ranges and were also to become the sites of future deposition.

The general uplift at the close of the Pliocene drove the sea out of the Great Valley and forced the Pacific shore farther seaward than it is now. The Merced embayment became dry land, and the exposed seafloor sediments were mildly folded. It is possible that the lower course of the Russian River, which drains directly westward into the sea, had its origin at this time. If so, it had to circumvent or maintain its course through the shallow folds. It is possible, too, that the drainage of the Great Valley escaped to the sea by following a devious course around the ends of the folds, but we have no direct evidence of this.

The Purisima Sea and the Priest Valley Strait likewise disappeared. Point Arena, Bodega Head, and the Point Reyes Peninsula, still south of their present positions, were all larger than now, part of a widened coastal strip including elevated sea floor. The Santa Lucia Range, too, was broader than now. No longer was there a seaway in the Salinas Valley separating the Santa Lucia Range from the Gabilan Range to the east. The Gabilan Range had by now attained about its present size,

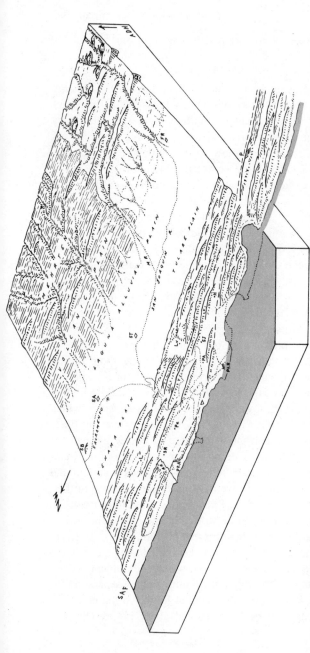

Figure 14. Late Pliocene mountain-making, 2 million years B.P. Uplift of the Sierra resulted in the Mountain Canyon Stage of valley development (see Figure 12). The Laguna Alluvial Plain was now dissected by streams which were opening out broad valleys below it. The region of the Coast Ranges was crumpled into broad folds, and the shoreline was driven some distance seaward. The inland sea no longer reached Middle California. Many basins were created in the Coast Ranges, including the Livermore Basin and a new basin west of present Clear Lake. (FR—Fresno; H—Hollister; L—Lower Lake; Lv—Livermore; MR—Merced River; P—Pinnacles; PA—Palo Alto; Pe—Petaluma; Pt.A—Point Arena; Pt.R—Point Reyes; RR—Russian River; SA—Sacramento; SAF—San Andreas Fault; SB—Sutter Buttes; SJ—San Jose; SR—Santa Rosa; ST—Stockton.)

largely due to renewed upheaval by faulting on the east side. The steepened gradients invigorated the Gabilan streams, thereby initiating the erosion that was to carve the Pinnacles volcanic rocks into the picturesque spires they display today.

To the north was a greatly expanded Santa Cruz Range. It, too, had been tilted up on its east side, but the western slope was crumpled locally into folds. The Point Reyes Peninsula was then at the latitude of San Jose, and Point Arena was just south of the Russian River.

Inland, interesting events were taking place. Northwest of Livermore (Figure 14), a pimplelike protuberance began to appear on the crest of one of the growing folds. The welt continued to grow until it measured about 5 miles (8 kilometers) across, at which time it ruptured. Out of the rupture rose a mass of ancient Franciscan rocks, including a body of dark intrusive igneous rock since altered to serpentine. So was Mount Diablo born.

The Sierra, during this episode of mountain-making, was tilted upward until it stood several thousand feet higher than formerly. The crest of the range may have stood about 7000 feet (2135 meters) above sea level at this time. The tilting, which had been cumulative during the latter part of the Pliocene, had two effects on the drainage of the Sierra. Whereas formerly some of the large Sierran rivers may have had their sources east of the present crest, the rapid uplift along the eastern scarp cut off their headward portions. Echo Summit, for example, a deep pass across the Sierran crest south of Lake Tahoe, was once the path of the South Fork of the American River, which entered from the east. Other low passes had similar origins. A second consequence of the tilting was the rejuvenation of the Sierran streams flowing down the western slopes. Erosion was accelerated, even in their lower courses. The Laguna Alluvial Plain beyond the foothills began to be trenched, and broad shallow valleys were opened out below it. The expansion of these valleys was to develop a widespread erosion surface below and beyond the alluvial plain.

Across the Great Valley, streams from the Coast Ranges

continued to deposit their sediment (Tehama and Tulare formations) along the margins of the valley. As the level of sediment rose, it progressively buried the lower slopes of the Coast Ranges and the eroded edges of older formations.

This was the setting at the advent of the Great Ice Age.

THE GREAT ICE AGE

The past 1.8 million years, known to geologists as the Pleistocene Epoch, was a time of climatic refrigeration when huge glaciers buried the northern half of North America at least four times and when mountainous areas were draped in valley glaciers. Although the Coast Ranges of Middle California were, with few exceptions, too low to house glaciers, the High Sierra was intensely glaciated and the landscape was radically altered. The foothills of the Sierra, the Great Valley, and the Coast Ranges were indirectly affected by glaciation and their landscapes were modified in numerous details.

At present, 6 million square miles (15.5 million square kilometers), or 10 percent of the land area of the Earth, are covered by glacial ice. The bulk of this ice is in Antarctica and Greenland, with highland ice caps and valley glaciers contributing the rest. During the Great Ice Age, glacial ice covered more than twice that area. A huge ice sheet, as large as that blanketing Antarctica today, and two or more miles thick, covered the northern half of North America as far south as the Ohio and Missouri rivers. A comparable ice sheet covered northern Europe and European Russia. At this time, the high crests of the Sierra Nevada were largely buried in ice. Great valley glaciers probed down the valleys of the eastern and western flanks of the range. Those on the west side descended to elevations of about 4000 feet (1220 meters), well short of the margin of the Great Valley. Long tongues of ice ground their way down Merced Valley on several occasions to create beautiful Yosemite Valley.

The Coast Ranges of Middle California supported valley glaciers in a few places. Snow Mountain, about 20 miles (32 kilometers) north of Clear Lake, was the only peak in our area to house glaciers, and these were very small features. Because this local glaciation probably took place in late Pleistocene time, we shall defer details until later. Farther north, however, toward the Trinity Alps, glaciation was more intense, as indicated by the glaciated topography.

In the Great Valley, the indirect effects of glaciation resulted from the floods of debris from the melting Sierran glaciers. In the Coast Ranges, on the other hand, the indirect effects were related to fluctuations in the level of the sea as water was extracted to form glacial ice and later returned to the sea as the ice melted. These fluctuations caused the sea at times to withdraw from the land and at other times to invade it. These activities affected the behavior of the Coast Range streams. When sea level stood low, streams were generally rejuvenated and eroded their valleys; when sea level stood high, their lower courses were submerged and deposition was encouraged. Stream behavior was also influenced by the fluctuating climates of the Ice Age, changing from cold and moist to warm and drier as glacial and interglacial stages alternated. And to add to the complexities, tectonic uplift and depression in the geologically unstable Coast Range region not only influenced stream behavior but acted independently to modify uplands and lowlands alike.

It is convenient to discuss the Pleistocene history of Middle California in two phases separated by a strong pulse of mountain-making. For convenience, we shall include the last 10,000 years in the Ice Age, in deference to the great expanses of the Earth still covered by ice. Technically, however, specialists separate the last 10,000 years as the Holocene or Recent Epoch.

PHASE I

1.8 to 1 million years B.P.

Coast Ranges. Even while the late Pliocene coastal mountains were rising, they were subject to attack by the invigorated streams on the steepening slopes. The sediments eroded from the mountains were washed down into the Pacific Ocean on the west, into downfolded and downfaulted basins among the ranges, and into the Great Valley on the east. Erosion of the

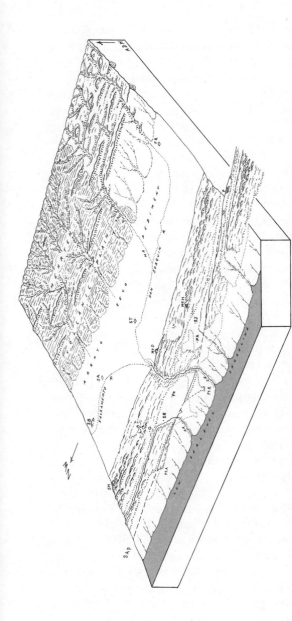

Figure 15. Great Ice Age—Phase I. The Pacific shoreline is shown during one of the glacial stages when much ocean water was locked up in great ice sheets on land. The drainage of the Great Valley now escaped seaward along approximately its present path. The basins and many of the valleys continued to serve as repositories of sediment, while the ridges were beveled except for residual mountain masses. The basin southwest of Lower Lake (L) became the site of intensive volcanic activity. The upper reaches of many Sierran valleys were now converted to glacial troughs. The Laguna Alluvial Plain was almost completely destroyed and replaced by a widespread erosion surface, the Arroyo Seco Pediment. (CR—California River; FR—Fresno; Lv—Livermore; MR—Merced River; Mt.D—Mount Diablo; Mt.H—Mount Hamilton; P—Pinnacles; PA—Palo Alto; Pe—Petaluma; Pt.A—Point Arena; Pt. R—Point Reyes; RR—Russian River; SA—Sacramento; SB—Sutter Buttes; SJ—San Jose; SM—Snow Mountain; SR—Santa Rosa; ST—Stockton.)

Coast Ranges continued even as the mountain-making forces weakened and died. The interior basins began to fill with terrestrial silts, sands, and gravels. The net result of erosion of the uplands and deposition in the lowlands was a widespread rolling plain surmounted by as yet unconsumed mountain masses (Figure 15). Most of these residual masses were the high interior portions of the former mountainous uplifts. That they were not completely consumed was due in large part to a new pulse of uplift in mid-Pleistocene time that reelevated the region and forced the streams to expend much of their energy in renewed excavation of their valleys. The unconsumed mountain residuals form the highest parts of the Santa Lucia Range, the Diablo Range, and the uplands north of San Francisco Bay.

During this early Pleistocene interval of erosion and deposition, the Livermore Basin, initiated during the late Pliocene mountain-making, was filled with a tremendous accumulation of fluvial gravels, sands, silts, and clays, with scattered lake beds. The deposits, known as the Livermore Gravels, are about 4000 feet (1220 meters) thick. They are thickest on the south side of the basin and the layers dip northward. On the north side of the basin they are buried under younger alluvium. It would appear, therefore, that the major source of the Livermore Gravels was the high central mass of the Diablo Range to the south. The enormous thickness of these fluvial deposits, with their floor far below sea level, indicates continuing subsidence of the basin during deposition. In the upper part of the formation there is a 10-foot (3 meter) thick tuff layer, evidence of volcanic activity somewhere in the general region.

Other gravelly deposits were laid down in a long narrow depression including the southern part of the San Francisco Bay lowland and extending far to the south past Hollister and along the west side of Gabilan Range. These largely fluvial deposits are known under different names in different places: on the west side of the San Francisco Bay lowland, they comprise the Santa Clara Formation, which may be about 3000 feet (900 meters) thick under the center of the lowland. As with the Livermore Gravels, the floor on which they rest is below sea level, so progressive subsidence must have taken place during deposition. The deposits are presently preserved as patches along the foot of the mountains bordering the San Francisco Bay lowland on the west. They are exposed in the banks of many of the streams that enter the lowland from the west, such as San Francisquito Creek that crosses the Stanford campus.

On the east side of the Bay lowland, particularly in the neighborhood of Irvington, equivalent coarse fluvial sediments, the Irvington Gravels, are preserved in a high, narrow terrace on which Mission San Jose is located. A number of large sand and gravel pits in the front of the terrace provide excellent exposures and have yielded an impressive collection of early Pleistocene fossil vertebrates, including camels, saber-toothed cats, wolves, and ancient horses. South of San Jose, the gravels include basaltic lava flows, a result of local volcanic activity.

Similar alluvial deposits, locally known as the San Benito Gravels after the town of that name just east of Pinnacles National Monument, were deposited in the long narrow lowland that extended north to the San Francisco Bay lowland. This narrow lowland roughly coincides with the San Andreas fault zone.

In the Salinas Valley between the Santa Lucia and Gabilan ranges, other gravels were deposited. These are the northern remnants of the Paso Robles Gravels, which have their greatest extent farther south in the vicinity of the town of that name. They extend over broad areas from the Salinas Valley to the Temblor Range east of the San Andreas Fault. The river which deposited the Plio-Pleistocene alluvial deposits in the Salinas River Valley entered the sea at Monterey Bay and spread silts and muds out on the sea floor. Drilling for oil around the margins of Monterey Bay reveals about 1500 feet (460 meters) of these sediments resting on marine Pliocene sediments.

The Plio-Pleistocene deposits of the Coast Ranges of Middle California were not everywhere deposited at the same time. In some places, the terrestrial deposits rest without interruption on Pliocene marine deposits, indicating continuous deposition as the Pliocene sea withdrew. In other places, the gravelly deposits rest on the eroded edges of Pliocene strata, indicating that deposition began only after a period of deformation and erosion of the earlier strata. Invariably, however, deposition of the Plio-Pleistocene gravels was terminated by a new episode of deformation in mid-Pleistocene time.

Meanwhile, the broad basin that had been occupied by the Merced Sea between the San Francisco Peninsula and the Russian River, and inland as far as the Sonoma Range, was now above water. Following a brief period of erosion, the marine deposits were covered by about 500 feet (150 meters) of terrestrial sediments. For the first time, there seems to have been a direct river outlet, rather than a marine strait, from the Great Valley to the sea at the approximate site of the Golden Gate. This is indicated by the first appearance of sediment from the Great Valley within actual river deposits in the upper part of the Merced Formation on the San Francisco Peninsula. For convenience, we shall refer to the river, the combined waters of the Sacramento and San Joaquin rivers, as the California River. The California River existed, of course, only when the Bay lowland was dry land, so that the interior drainage could extend its course across it. The outlet area at this time was low; there was no Golden Gate as yet. Exactly where the drainage escaped seaward from the Great Valley immediately prior to this time is not known. It will be remembered that at many times in the past, the Great Valley drained directly into embayments of the sea that extended into the valley. These exits, however, were cut off in the Pliocene (see Figure 13). At that time, the Great Valley drainage probably escaped to the sea far to the south through other openings in the Coast Ranges. Subsequent elevation of the southern Coast Ranges may have closed off the southern routes, forcing the interior drainage back to the San Francisco Bay region.

We have already noted that the present westerly bend in the lower course of the Russian River was probably related to the withdrawal of the Merced Sea from its embayment (Figures 13 and 14).

The Cache Creek Basin, east of present Clear Lake, still continued to receive sediment in early Pleistocene time, but the new basin to the west, created by faulting at the close of the

Pliocene, now became the scene of other activity. This fault basin, instead of filling with sediment, became the site of volcanic accumulation. Even before deposition of the Cache Creek Basin deposits ceased, volcanoes began to erupt here, spewing lava and explosion products over an area of about 150 square miles (390 square kilometers). The activity was intermittent; at times, active volcanoes dotted the landscape; at other times, the activity died down and the volcanoes were eroded. In places, sticky lavas piled up in mushroomlike domes. But Mount Konocti and many of the small cinder cones in the area had not yet made their appearance.

Whether a forerunner of Clear Lake existed at any time during the early Pleistocene volcanic phase is not yet known; drilling into the bottom sediments has not reached the base of the deposits. It seems probable, however, that a lake existed here at various intervals during early Pleistocene time.

It is interesting to note that the major centers of volcanic activity in the Coast Ranges have been shifting north since late Oligocene time. In the Miocene Epoch (see Figure 10), the main centers of eruption were near Hollister at about the latitude of Monterey; in the Pliocene Epoch, the major center was in the Sonoma Range just north of San Francisco Bay; and in the Pleistocene Epoch, it moved north to the vicinity of Clear Lake. Some believe that this traveling hot spot represents the junction of three crustal plates that is being carried north below the edge of the continent with the moving Pacific Plate.

Great Valley. At the same time that the late Pliocene Coast Ranges were being eroded and their sediment was filling interior valleys and basins, other sediments were being spread eastward into the Great Valley, continuing the expansion of the Tehama and Tulare plains. But it was along the Sierra side of the Great Valley that more intriguing events were taking place. The continued tilting of the Sierra kept invigorating the west-flowing streams, causing them to deepen their valleys not only in the mountains but for some distance out on the valley floor. Thus, the Laguna Alluvial Plain was trenched by the merging streams, which then began to open out their valley floors to great widths. The expansion of this lower plain resulted in the destruction of much of the higher Laguna Plain. The lower plain, an erosional rather than a depositional feature, is known as the Arroyo Seco Pediment. Although the name Arroyo Seco Pediment was originally applied to an area in the Sacramento Valley, we will, for convenience, apply it to all equivalents of this surface along the eastern margin of the Great Valley.

Sierra Nevada. Within the mountains, erosion was the rule. The rivers continued to deepen their canyons, and as they did so they invigorated their tributaries coming in from the sides. As a result, the tributaries, like expanding gullies, spread their headward tentacles farther and farther into the backslope of the range to convert it into an intricate network of deep valleys. In the area of the Mehrten Volcanic Plain (Photo 12), however, enough of the flat interstream divides remain to clearly reveal the original smooth westward slope.

Because of repeated uplift, the Sierra Nevada was now high enough to act as an exceedingly efficient generator of its own precipitation. Although moisture-laden winds coming off the Pacific dropped some of their moisture on the Coast Ranges, most of it was retained to fall out as precipitation during the long ascent of the Sierra. By the time the winds reached the crest, almost all of the moisture had been wrung out of them and little remained for the parched areas to the east.

It was at this time, too, nearly 2 million years ago, that the climate began to turn cold, and each year more and more of the precipitation was in the form of snow, and more and more of this snow persisted through the shortening summers. At first, only the highest parts of the Sierra were covered by permanent snow; but as the climate continued to chill, the snowline crept slowly down to lower levels. Snow accumulated to greatest thicknesses in the sheltered heads of valleys. Part of the snow in the valley heads was direct snowfall, but much was snow that the wind swept from exposed summits into the valleys. Because the winds in these latitudes are primarily from the west, most of the windswept snow came to rest in the heads of the valleys draining the east slope of the range. Eventually, the snow became so thick that its weight squeezed and compacted the lower levels into solid ice. As the weight continued to increase, the deeper ice was forced out of the valley heads and moved downslope carrying the overlying snow with it. Thus were valley glaciers born. As they moved away from their source, they carried with them blocks of rock pried from the enclosing slopes as well as debris abraded from their floors and sides. Through such activity, the valley heads were converted to huge rocky amphitheaters known as *cirques*. As the glaciers moved downvalley, they steepened and broadened the valleys, converting their cross-sectional profiles from the original river-eroded V-shapes to much more impressive U-shaped troughs. Yosemite Valley is a classic example. It was the deepening and widening of the main valleys that left the tributary valleys hanging high overhead so that they were forced to enter the main valley by way of waterfalls. Bridal Veil Falls in Yosemite is an example well known to visitors to the region.

The debris carried downvalley by the valley glaciers was dumped at the melting edge to form hummocky ridges known as *end moraines*. When the glacier front later melted back, additional debris was spread in its wake as *ground moraine*. And when, as often happened, the ice front halted for a considerable time during retreat, another ridge was built up at the ice front, a *recessional moraine*.

Beyond the glacier fronts in the western valleys of the Sierra, meltwaters carried much sediment *(outwash)* far downstream and into the Great Valley. So it was that, during wasting of the glaciers, the floors of the Sierran valleys in the foothills and beyond became clogged with debris. During subsequent expansion of glaciers, when outwash was at a minimum, the same streams incised their channels into the previously deposited outwash floodplains. When these glaciers in turn began to waste away, a new flood of meltwater debris was flumed downvalley. In the foothills, this debris was restricted to the incised channels, but beyond the foothills, where the valleys were increasingly shallow, the sediment filled them to overflowing and spread widely from one valley to the next. Thus, the recurrent Sierran glaciations are recorded in the foothill valleys by a succession of progressively lower and younger terraces of outwash. Each terrace represents the top of an outwash floodplain in an inner valley eroded during the prior interglacial interval (Figure 16). But farther out in the Great Valley, the effects of interglacial erosion were negligible, and each succeeding flood of outwash buried the preceding deposits with little evidence of interruption.

In the absence of glaciers, the cycles of erosion and deposition in the Coast Ranges and along the west side of the Great Valley were not related to glaciation and did not correlate directly with the erosional and depositional record on the Sierran side. Streams draining into the Great Valley from the Coast Ranges continued to deposit their sediment more or less uninterruptedly. Such interruptions as did occur were primarily caused by climatic changes.

The behavior of the Coast Range streams that drained into the Pacific Ocean was determined primarily by fluctuations in sea level, by climatic changes, and locally by tectonic upheaval. The sea level changes were, of course, indirectly related to glaciation. During the glacial stages, not only was sea level lower but the climate was cooler and wetter. During the interglacial stages, when sea level stood high, the climate was warmer and drier. When sea level stood low, the Pacific streams deepened their valleys. As sea level rose during the interglacial stages, the valleys were filled with alluvial and deltaic sediments. In the Sierra, on the other hand, deposition took place during the glacial stages, and erosion during the

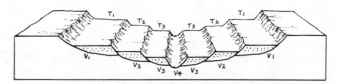

Figure 16. Glacial terraces in Sierran foothill valleys. V_1 represents the floor of the highest buried valley, which was filled with sediment to the level T_1. This broad valley was trenched by an inner valley V_2, which was subsequently filled to the level T_2. Similarly for V_3 and T_3. The last stage of incision is shown by the valley V_4. Such terraces record episodes of glaciation and deglaciation in the high Sierra.

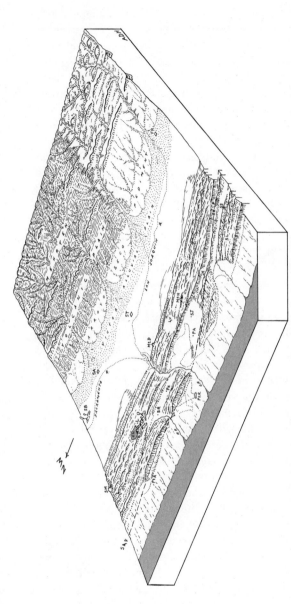

Figure 17. Mid–Ice Age mountain-making, 1 million years B.P. The major topographic units of the Coast Ranges were blocked out at this time, mainly by faulting. The Arroyo Seco Pediment was now being dissected, and its debris was added to that from the Sierra to form the Victor Alluvial Plain. The landscape elements west of the San Andreas Fault were now close to their present-day positions. (CL—Clear Lake; CR—California River; FR—Fresno; GR—Gabilan Range; H—Hollister; L—Lower Lake; Lv—Livermore; Mt.D—Mount Diablo; Mt.H—Mount Hamilton; P—Pinnacles; PA—Palo Alto; Pe—Petaluma; Pt.A—Point Arena; Pt.R—Point Reyes; RR—Russian River; SA—Sacramento; SB—Sutter Buttes; SJ—San Jose; SL—Santa Lucia Range; SM—Snow Mountain; SR—Santa Rosa; SS—Sierra de Salinas; ST—Stockton.)

interglacial stages. In short, the cycles of erosion and deposition in the two regions were out of phase.

Local uplifts in the Coast Ranges added additional complexities to the patterns described above. For example, rapid uplift locally induced stream erosion in spite of a slowly rising sea level. In this situation, both the land and the sea were rising, but because the land was rising faster, the net result was acceleration of stream gradients and erosion. Rapid crustal depression, on the other hand, caused deposition in some places in spite of a lowering sea level because the land was sinking faster than sea level. The net result was a drowning and filling of valleys. The environs of Monterey provide a good example of these phenomena, but because the record dates largely from the late Pleistocene, we will defer the discussion until later.

MID–ICE AGE MOUNTAIN–MAKING

Deposition of the widespread Plio-Pleistocene sediments in the Coast Ranges was followed by yet another strong pulse of mountain-making (Figure 17). This pulse differed from those described earlier in that faulting played a dominant role. The entire Coast Range belt of Middle California began to break up into blocks, some of which rose while others subsided. Many of the rising blocks were tilted in trapdoor fashion. Many of those near the coast present steep scarps to the east and gentle slopes to the west like miniature replicas of the Sierra Nevada.

The Santa Lucia Range, by now only 10 to 15 miles (16 to 24 kilometers) south of its present position, was bodily uplifted along faults on both the inland and seaward sides, and other faults crisscrossed the interior of the range. The uplift was intermittent, as indicated by remnants of marine terraces high up on the coastal flanks, suggesting long pauses in uplift during which the sea deeply notched the rising slopes. Even more interesting was the dislocation of a long narrow segment of the eastern flank of the range. This segment rose along a great fracture on its east side and dropped along a fault on the west, a bold eastward-facing trapdoor we know as the Sierra de Salinas.

Eastward across the Salinas Valley from the Santa Lucia Range, the granitic Gabilan Range was elevated still further, slightly higher on the east along the San Andreas Fault. At the same time, the low strip of erosional plain extending south from the Gabilan Range was also tilted westward. At its north end, the tilt block split into two linear trapdoors, a narrow eastern one which bears the name Mustang Ridge, and a much broader one to the west which is known as Gabilan Mesa. It is interesting that the valleys eroded into the western slope of Gabilan Mesa are asymmetric in cross-profile—that is, the north-facing slopes are steeper than the slopes facing south. This is because the north-facing slopes are shadier and moister and are protected by a cover of vegetation, while the south-facing slopes are unprotected and therefore more rapidly worn down by erosion.

North of Monterey Bay, the Santa Cruz Range was also uplifted. It apparently started out as another trapdoor, with uplift greatest along the San Andreas Fault on the east; but during upheaval its back was broken in several places, giving rise to the lesser trapdoors of Montara Mountain, Butano Ridge, and Ben Lomond Mountain. Across the San Andreas Fault and slightly farther north, the isolated tilt block of San Bruno Mountain rose above the lowlands, but with the fault scarp facing southwest (Photo 11). The downdropped area southwest of the scarp is the Merced Lowland, extending southeastward from the Pacific coast at Lake Merced to the San Francisco lowland at South San Francisco. In this down-dropped lowland are preserved several thousand feet of the Plio-Pleistocene Merced sediments described earlier.

Across the long lowland now occupied in part by San Francisco Bay, the entire Diablo Range from San Pablo Bay in the north to the southern limit of our area was bodily uplifted, presenting a bold scarp overlooking the San Francisco Bay lowland. The range was broken by many crossfaults between which segments were tilted up either to the north or south. And some of the intermontane basins, such as the Livermore Basin, were further depressed, warping the earlier Livermore Gravels.

The area between the opposing scarps of the Santa Cruz and

Diablo ranges foundered into the depths, breaking up in the process. Most of the resulting long narrow blocks are now completely buried under sediment washed in from the surrounding mountains. The top of one of these blocks, not quite completely buried, sticks up above the surrounding sediments at the south end of San Francisco Bay. The eroded crest forms the Coyote Hills at the east end of the Dumbarton Bridge causeway.

Faulting on a grand scale also took place north of San Francisco Bay. Part of the earlier Merced plain was uplifted and broken to form the Marin Mountains. Subsequent erosion has removed the Merced sediments from this uplifted block. The northern part of the Merced plain, however, was elevated only slightly and today forms a broad, low saddle between the Marin Mountains on the south and the Mendocino Range on the north.

West of the Marin Mountains, across the San Andreas Fault, the Point Reyes Peninsula was again elevated and its broad base along the fault valley may have been tilted up as another trapdoor to form Inverness Ridge. The Farallon Islands, far off the coast, probably represent the eroded crest of still another trapdoor uplift (Photo 13).

The Sonoma Range, east of the Santa Rosa–Petaluma lowland, was uplifted along the northern extension of the Hayward Fault, the same fault that follows the foot of the Diablo Range south of San Francisco Bay. It is a matter of some concern that this active fault passes directly under the football stadium of the University of California and crosses the paths of several water aqueducts.

From the Sonoma range to the Great Valley, the entire Coast Range belt was broken by major faults, now followed by broad linear valleys that separate the northwest-trending ranges.

As a result of the uplift and tilting, the Plio-Pleistocene gravels described earlier were upturned, folded, and broken close to the active faults. In one exposure on the Stanford campus, beds of the Santa Clara gravels may be seen standing at an angle of 70 degrees. The widespread vertical faulting does not mean that lateral displacement had ceased along the

San Andreas and similar faults. The Santa Clara gravels themselves appear to have been offset 16 to 18 miles (26 to 29 kilometers) since they were deposited.

Along the margin of the Great Valley, the Tehama and Tulare Formations were flexed upward along the flank of the Coast Ranges and also crumpled into long, low welts extending out into the valley. The Montezuma Hills, within the big bend of the Sacramento River where it turns west toward San Francisco Bay, are composed of Pleistocene sediments raised into a circular dome.

Another result of the general uplift of the Coast Ranges was the rejuvenation of the streams and accelerated erosion. Valleys were deepened in the rising blocks, the crests of which still preserve remnants of the lowland plain from which they were elevated. Whether the streams that drained into the Pacific eroded or deposited depended in part on the level of the sea at the time. But streams draining into the Great Valley continued to dump their debris on the valley floor.

Meanwhile, volcanism continued in the Clear Lake volcanic field, probably accelerating as deep new fractures penetrated the crust. A series of closely spaced fractures followed the valley of Big Sulphur Creek. Heated vapors from the deep body of magma converted much of the ground water above to steam, and many steam vents sprang up along the faults. In this way the geysers north of Healdsburg came into existence. Actually, the steam vents are not geysers; they do not erupt like Old Faithful in Yellowstone National Park and are more properly called fumaroles or gas vents.

I do not mean to imply that the upheaval of crustal blocks all over the Coast Ranges was a catastrophic event. The rate of uplift was actually so slow that the California River was able to maintain its path to the sea whose shoreline lay some distance beyond the Golden Gate. There was no San Francisco Bay at that time, and there is no record of sediments this old under the waters of San Francisco Bay. The Bay lowland was then apparently dry land. It is likely, however, that at times the crustal blocks rose rapidly enough to create low dams across California River and form shallow temporary lakes. The overflow waters, however, continued to spill through the slightly elevated channels and flow on to the distant shore. As the shallow temporary lakes became filled with sediment, the course of the California River once more became unified, uninterrupted by intermittent ponds.

Across the Great Valley, the great trapdoor of the Sierra Nevada continued to rise, higher in the south than in the north. Each pulse of uplift invigorated the west-flowing streams, causing them to deepen their valleys. That there was little other deformation is indicated by the relatively undisturbed broad, flat, westward-sloping divides between the major valleys (see Photo 12). However, some small vertical offsets of 100 feet (30 meters) or so occurred along northwest-trending faults in the foothills. In each case, the ground immediately west of the fault was tilted up.

It has been estimated that the gradients of the valleys of the Yuba and Mokelumne rivers in the northern Sierra were steepened as much as 60 to 70 feet per mile (18 to 21 meters per kilometer) since tilting of the Sierra began. This was, of course, in addition to the gradients that prevailed on the western slope of the range prior to tilting. In the southern Sierra, south of the Mehrten Volcanic Plain, the evidence for magnitude of uplift is less direct, and estimates vary considerably. It appears that whereas the northern Sierra rose dominantly by a trapdoor-like tilting, this tilting gave way southward to more nearly vertical uplift between major faults on both sides of the range. There are no smooth, flat, gently sloping divides between the major valleys. Instead, the west flank of the southern Sierra seems to consist of several vast terraces stepping down toward the Great Valley. Some geologists regard the steps as fault-controlled, each segment on the west having been dropped down relative to that on the east. Others believe that the steps are lowland erosion levels that were uplifted with the range. Still others believe that the steps reflect different rates of erosion on slightly different granitic rocks that make up the multiple Sierra Nevada batholith. In any event, whatever the style of uplift, the Sierra underwent a strong surge of uplift at the same time that the Coast Ranges to the west were cracking up.

The renewed uplift of the Sierra again rejuvenated the west-flowing rivers, resulting in renewed erosion of their valleys. On entering the Great Valley, these streams trenched the Arroyo Seco Pediment, opening broad shallow valleys below it. The debris resulting from this erosion was spread out beyond the margin of the pediment as alluvial fans. These eventually coalesced to form a widespread depositional plain, the Victor Alluvial Plain.

PHASE II

The last million years

During the last million years, the great fault scarps formed during the mid–Ice Age mountain-making were weathered and eroded and their sharp edges blunted (see Figure 1). And the rejuvenated streams that rushed down the slopes of the uplifted blocks eroded a complex of deep, narrow valleys to create the present rugged landscape. Only where the streams have not yet sent their tentacles into the heart of the ranges are broad level remnants of the earlier lowland plain preserved.

The rocks of the Coast Ranges, because of the numerous mountain-making episodes they have undergone, are no longer horizontal; they dip at all angles up to vertical. The rocks vary considerably in their resistance to erosion: some are durable and stand up as ridges; others are weak and have been eroded to valleys. And the crushed rock along major faults provided especially weak zones along which have been excavated major valleys.

We noted that during the first phase of the Ice Age much of the landscape, weak and durable rock alike, had been worn down to a low plain close to sea level (see Figure 15). Sea level is the ultimate level to which streams can erode their valleys, no matter how weak the rocks may be. Thus, even weak rocks cease to be lowered as they reach the level of the sea, while the higher, more resistant rocks continue to be lowered. Hence, erosional plains close to sea level truncate weak and resistant rocks almost indiscriminately. But the mid–Ice Age orogeny elevated large areas of the previously formed plain, again providing the streams with steep gradients. With their newfound vigor, the streams once again began to attack the uplifted segments of the plain, eroding the weaker rocks more rapidly than the resistant ones. The crush zones along major faults were etched out to form many long, straight valleys. A classic example of this is the valley, or rather the series of aligned valleys that were etched out along the San Andreas Fault. South of San Francisco in the Santa Cruz Range, the crush zone has been eroded into a long, straight valley now occupied in part by San Andreas and Crystal Spring lakes (see Photo 11). The valley here is more than half a mile (0.8 kilometer) wide. The material in the crush zone can be examined firsthand in the road cuts a short distance up the grade from the east end of the causeway crossing Crystal Springs Lake. Incidentally, at the time of the San Francisco earthquake of 1906, this causeway was displaced about 7 feet (2.1 meters), the west side having moved north.

North of the bay, the crushed material along the fault was eroded to form the remarkably straight valley occupied in the south by Bolinas Lagoon and farther north by Tomales Bay. In this area, the ground was displaced up to 21 feet (6.4 meters) in 1906. With interruptions, the line of the San Andreas Fault can be followed by its topographic expression for more than 600 miles (970 kilometers).

Not all the major valleys are stream-eroded. As we have seen, the San Francisco Bay–Santa Clara lowland is a graben that foundered between major faults on both sides of the valley. It is similar in origin to the basin of the Dead Sea between Israel and Jordan, and to the Rhine Valley between the Vosges Mountains on the west and the Black Forest on the east. Other examples are the "Rift Valleys" of East Africa, occupied by Lakes Nyassa, Tanganyika, and others.

The end result of the erosion of the belts of weak rock, of the foundering of graben, and of the continued downtilting of other blocks was the present Coast Range landscape with its parallel ranges and valleys approximately parallel to the coast.

The elongate dome of Mount Diablo was also differentially eroded. The weaker rocks were etched out, leaving the resistant layers standing as ridges encircling the central core (see Photo 3).

The effects of differential erosion were similar throughout the Coast Ranges, with ridges and valleys alternating. Only in those areas where all the layered rocks had been removed from above the hard, more or less uniformly resistant ancient rocks was there no opportunity for selective etching of the landscape. In these areas, except for valleys etched along faults, there is little parallelism of ridges and valleys. The Mendocino Range, the Marin Mountains, and parts of the Diablo Range south of the Livermore Basin are examples of such irregularly dissected areas.

It will also be recalled that far to the south, in the Gabilan Range, a mass of volcanic rocks had earlier been dropped down into the older rocks and thereby preserved from removal during the early Pleistocene erosion. Now, thanks to the most recent upheaval, invigorated streams attacked these weak and fractured volcanic rocks, etching them into a filigree of spires and pinnacles of such unusual beauty that the area has been set aside as the Pinnacles National Monument (Photo 14).

Meanwhile, in the Great Valley, floods of debris continued to pour down the Sierran valleys during the melting of the Sierran glaciers. The outwash floodplains so formed were trenched in the interglacial episodes when glaciers and outwash were absent. Thus, each valley within the foothills and for a few miles beyond displays a succession of terraces. For a few miles beyond the mountain front, these terraced valleys lie below the level of the Arroyo Seco Pediment. Beyond the Arroyo Seco Pediment, the sediment was spread out as an alluvial plain which increased in size with each episode of glacier thaw. In Figure 17, this alluvial plain, actually consisting of at least two ages of plains, is referred to as the Victor Alluvial Plain.

On the west side of the Great Valley, following the Mid–Ice Age mountain-making, the alternations of erosion and sedimentation were primarily related to the fluctuating climates of the Ice Age; there were only a few small glaciers in the Coast Ranges of middle California. And the terraces within the valleys that drained into the Pacific were due to episodes of erosion and deposition related to fluctuations in both sea level and climate. In the Salinas Valley, the highest river terrace probably dates from the last interglacial stage, about 100,000 years ago. The surface of the terrace was part of the floodplain of the Salinas River when the river emptied into the sea 100 feet (30 meters) above present sea level. During subsequent drops in sea level, the Salinas River eroded its valley below present sea level. Since the waning of the last glaciers and the subsequent rise in sea level, the sea invaded the deepened valley of the Salinas River and, about 6000 years ago, penetrated as far inland as the city of Salinas. The marine—or rather, estuarine—sediments that record this advance are now buried under delta and floodplain sediments, the accumulation of which drove the sea from the interior.

The Monterey Bay lowland continued to subside between the rising mountains to the north and south. Deposition in the subsiding basin was not continuous, however. It was interrupted by minor cycles of erosion and deposition due to fluctuations in sea level and climate. The record has been investigated most thoroughly in the northern part of the lowland, in the region of Watsonville, but comparable events probably took place to the south. During each glacial stage, when sea level stood lower, the mouth of the Pajaro River was far seaward of the present shoreline and hundreds of feet lower. At such times the river was able to erode its valley hundreds of feet deeper than now. As the glaciers melted and sea level rose, the lower course of the Pajaro River became an estuary and the estuarine conditions advanced landward as sea level continued to rise. When the sea stood highest, the estuary extended far inland and was eventually filled with sediment from the upper reaches of the valley. In brief, valley erosion took place during glacial lowerings of sea level, and deposition followed during rising sea level. The net result of this in the Monterey Bay lowland was a succession of terraces, the surfaces of which record high stands of the sea, and the trenches below them, low stands. The sequence of events in the Watsonville area is especially interesting when considered in relation to the effects of rising and falling sea level on the upland areas to the north and south of the basin. At each high stand of the sea, when the estuaries within the Monterey basin were being filled with sediment, the sea was eroding the adjacent steep coastal slopes to form broad submarine platforms extending to the foot of prominent sea cliffs. During lowering sea levels, when the previously deposited valley fills of the Monterey basin were being trenched to form terraces, the marine platforms on the slopes of the nearby mountains were being exposed and covered with sediment from the slopes behind.

In the transitional zones between the subsiding Monterey basin and the mountains on either side, some of the terraces and some of the marginal basin deposits are flexed down toward the deeper parts of the basin. The same subsidence, coupled with tectonic uplift of the mountains to the north, caused a southward shifting of the courses of the streams draining into Monterey Bay. A succession of old channels remains as a record of their former positions. The Pajaro River itself once entered the Monterey basin at a different site, but the point of entry has been shifted along the San Andreas Fault.

At some time during the displacement along the San Andreas Fault, probably between 10,000 and 5,000 years B.P., landslides are believed to have blocked Pajaro River and to have created a large lake east of the fault to which the name Lake San Benito has been applied. This temporary lake, with its surface 300 feet (90 meters) above sea level, flooded the Santa Clara Valley as far north as Morgan Hill, where it spilled over a low divide to empty into San Francisco Bay. It extended south beyond Hollister, giving a total length of about 30 miles (48 kilometers) and a maximum width of about 12 miles (19 kilometers). The lake level is recorded by deltas and terraces.

Equally interesting events were also taking place offshore in Monterey Bay. The submarine canyon we noted earlier is crossed by several major faults of the San Andreas type. Lateral movement along these faults has periodically displaced the mouth of the canyon to the north, requiring submarine erosion of a new and more direct lower course to the sea floor. The newer courses were in turn displaced northward. It is believed that the displacements began millions of years ago, but the present lower course is only about 1 million years old.

Crustal movements did not cease with the episode of mountain-making shown in Figure 17. On the contrary, they are still going on, as attested to by the many earthquakes that shake the region, by ground offsets as in the earthquake of 1906, by the presence of late Pleistocene coastal terraces too high to be accounted for by interglacial high stands of the sea, and by warping of even the youngest terraces. We mentioned the Montezuma Hills in the big bend of the Sacramento River. Its domelike rise is probably responsible for the deflection of the Sacramento River around it. The dome may still be rising. Crustal instability is also attested to by precise surveys across major faults. Present activity along the San Andreas Fault is warping the ground for some distance on either side. Eventually, a displacement will occur, but exactly where or when is presently impossible to say, although progress is being made. Measurements between Mount Diablo and the Farallon Islands indicate that the sea floor from which the islands rise is moving northwest almost a half inch (1.25 centimeters) per year. The shift has totaled 5 feet (1.5 meters) since the earthquake of 1906.

Volcanic activity, which had begun about 2 million years ago in the Clear Lake region, continued on a grand scale throughout the Ice Age. Volcanic cones rose sporadically, pouring out flows of light and dark lavas and spewing great quantities of ash, cinders, and rock fragments over the countryside. The rhyolite lavas of Cobb Mountain just east of the Geysers steam field were extruded about one million years ago. Mount Konocti, 3000 feet (915 meters) high, on the west side of Clear Lake (Photo 15), is built of lavas and cinders piled up over a span of 150,000 years and terminating about 250,000 years ago. During the height of the volcanic activity, several huge lava flows advanced toward each other just southwest of Clear Lake but failed to make contact with each other. The deep depression between them is the present site of Thurston Lake.

Sediments cored from beneath the floor of Clear Lake indicate that the lake has been in existence for at least the last 135,000 years. Deeper drilling will probably extend its age farther back into the past. In its early stages, Clear Lake probably drained northwest from Lakeport into Scotts Creek and then past Blue Lakes and down Cold Creek to the Russian River (Figure 18). The upper headwater portion of Scotts Creek flowed east into the lake near Lakeport. Excessive deposition by upper Scotts Creek during the last glacial stage resulted in construction of a large alluvial fan over which the creek periodically changed course. The increasing height of the fan kept raising the level of the lake outlet until the lake was stabilized at the 1600-foot (490-meter) level, almost 300 feet (90 meters) above present lake level. This high stand of the

Figure 18. Oblique view of Clear Lake region. The lines of latitude and longitude are at 5-minute intervals. The distance between successive parallels is slightly less than 6 miles (10 kilometers).

lake is recorded by both wave-eroded platforms and deltaic terraces. At one time during this high stand, when upper Scotts Creek happened to be flowing down the north side of its alluvial fan, it spilled over into the outlet stream somewhere in the vicinity of present Scotts Valley. By deepening its valley to the level of the outlet stream it fixed itself in this course and was no longer able to enter Clear Lake directly. Scotts Valley was rapidly opened out in the loose sediments of the Lakeport Plain while the valley of Scotts Creek to the north remained constricted in harder rocks, The present eastern outlet of the lake was probably initiated when Cache Creek, flowing into the Sacramento Valley, expanded headward through the weak Cache Creek sediments until it tapped Clear Lake. Erosion along invigorated Cache Creek lowered the lake to its present level. The lowering, however, was intermittent, as indicated by terraces below the 1600-foot level. These intermittent halts in the subsiding lake level may have been due to retarded downcutting by Cache Creek as it encountered resistant layers of rock along its outlet course. In more recent times, the large landslides at Blue Lakes created a minor drainage change on the northwest side of the Clear Lake Basin. Prior to the landslides, all of Scotts Creek south of Blue Lakes drained into Cold Creek and on down into the Russian River to the west. The landslides blocked Cold Creek and impounded a separate lake, Greater Blue Lake, which extended back into Scotts Valley but did not connect with Clear Lake. At its highest, Greater Blue Lake spilled across a low place in the ridge on the east and continued into the broad plain at the north end of Clear Lake occupied by the town of Upper Lake. It then turned south into Clear Lake. Scotts Creek has maintained this roundabout course ever since.

Across Clear Lake from Mount Konocti lies Borax Lake with an interesting history of its own. It occupies a former embayment of Clear Lake which was cut off from the lake proper by a lava flow about 90,000 years ago. The lava is dark and glassy (*obsidian*) with a light frothy top (*pumice*). The obsidian was widely used by the Indians of the area for arrowheads and tools. The borax which accumulated in the basin was mined here in 1865 before the discovery of the huge deposits in Death Valley.

Relatively recent volcanic activity is attested to by the fresh cinder cone and flow at Buckingham Peak on the north side of Mt. Konocti, and by Round Mountain, about 2.5 miles (4 kilometers) northeast of Clearlake Oaks, another fresh cinder cone, but marred by quarrying operations. A lava flow at Sulphur Bank, on the south side of the east arm of Clear Lake, has been dated as 45,000 years old.

The last eruptions, the evidence for which consists of ash beds in the cores from below Clear Lake and the creation of the crater of Little Borax Lake at the northeast base of Mt. Konocti, took place only 10,000 years ago.

The Clear Lake volcanic field is bounded and crossed by faults, and is still being depressed to the northeast. As a result, the lacustrine and fluvial deposits dip to the north and the deepest water is in that direction.

That the Clear Lake region is still highly unstable is indicated by the numerous earthquakes that beset the region. There is no reason to believe that volcanic activity may not resume at some time. We have already noted that a body of magma probably lies only 6 miles (9.5 kilometers) or so below the surface. This accounts for the many fumaroles and hot springs in the region. Hot vapors issuing from the ground may be seen at Borax Lake and Sulphur Bank.

The geysers steam area, about 15 miles (24 kilometers) south of Clear Lake, is now a major source of geothermal energy (Photo 16). The first producing well was drilled in 1955. The field includes the world's most prolific geothermal steam well, which, when drilled in 1971, flowed superheated steam at a rate of 386,000 pounds (175,000 kilograms) per hour, energy sufficient to satisfy a community of almost 20,000 people. The deepest well at present is more than 8000 feet (2440 meters) deep.

As for San Francisco Bay, it is a relatively recent feature of the landscape. We noted earlier that even in early Pleistocene time, before the mid–Ice Age mountain-making, the area of present-day San Francisco Bay was dry land mantled by the

93

upper, terrestrial beds of the Merced Formation. The California River emptied into the sea somewhere outside the Golden Gate. Following the mid–Ice Age orogeny, the California River deeply eroded its valley, probably during the last glacial stage when sea level stood very low. When the last glaciers began to melt away, the Pacific shoreline was probably far out toward the edge of the continental shelf, 15 to 20 miles (24 to 32 kilometers) off the present coast. As sea level rose, it encroached over the exposed shelf and a long estuary projected ahead into the mouth of the California River. About 10,000 years ago, the sea penetrated the Golden Gate and spread inland between the confining mountain blocks. It reached Richardson Bay near Sausalito about 8700 years ago. The beginnings of San Francisco Bay approximate the time of widespread Indian habitation of the United States. Indian shell heaps are now found below sea level at Emeryville near Berkeley and elsewhere around the bay. Indian bones found in the banks of San Francisquito Creek on the Stanford campus have been dated at about 5000 years. And, at about the same time, an Indian met a sudden death, probably by drowning, along a tidewater stream entering the sea at the site of the present San Francisco Civic Center subway station. The bones, in silts containing both freshwater and marine organisms, were uncovered 26 feet (8 meters) below present sea level. In West Berkeley, a refuse shell heap presumably constructed along the ancient shoreline is dated at 2700 years.

The bay itself is underlain by thick muds. These apparently accumulated in the deeper parts of the bay during melting of the last glaciers. Where the water was clear, however, oyster beds flourished. The most widespread of these accumulated between 2300 and 2500 years ago. Smaller deposits as old as 5700 years have been discovered in deeper drilling, and patchy beds have formed up to recent times. The main beds are now exploited for the manufacture of cement.

An interesting side effect of the rising sea level was the local invasion of the San Andreas Fault valley. The fault valley parallels Inverness Ridge on the east side of the Point Reyes Peninsula. The sea invaded both ends of the valley to form

Bolinas Lagoon on the south and Tomales Bay on the north. The latter bay extends 15 miles (24 kilometers) south into the valley.

The marine terraces carved into the flanks of the coastal mountains imply either that the sea once rose to those heights or that the land has been raised to those elevations. Probably both events took place. But the high terraces, well out of reach of the interglacial seas, had to have been uplifted to those heights.

The origin of the marine terraces becomes clear if we examine what is going on along the coast today. The waves are battering the shore, driving a sea cliff inland and leaving behind a smooth rock surface under water. Should sea level drop or the land rise, we would see, out of water, a broad flat, like the broad terrace followed by the coast highway south of Half Moon Bay or along the Sonoma coast, or like those etched into the mountain slopes up to a thousand feet (300 meters) above sea level.

The sea has also busily engaged itself in the construction of beaches, sand spits, and bars along the coast. The main oceanic current along this part of the Pacific Coast is toward the south, but irregularities of the shore create countereddies which transport sand northward. One of these is responsible for the northward drift of sand, derived from cliffs many miles south of the Golden Gate, that has led to the fashioning of Ocean Beach at Golden Gate Park. Northerly currents have also built bars across the mouths of many of the valleys that were drowned during the flooding of this area. Merced Lake, now in part artifically contained, occupies one of these valleys shut off from the sea. Many smaller landlocked bays have been completely filled with sediment and now appear as low, flat, often swampy areas behind sand ridges.

To the north of Golden Gate, the drifting sands were carried across the mouth of Bolinas Lagoon, nearly isolating it from the ocean. On the south side of Pt. Reyes Peninsula, the former valley system now occupied by Drakes Bay was also isolated from the sea by the growth of sand spits.

Before the San Francisco Peninsula was densely settled, the

vigorous winds swept the sands of Ocean Beach inland, creating extensive fields of sand dunes. The reclamation of a large part of this dune field in the creation of Golden Gate Park was a notable achievement.

Few of us are probably aware of the presence of a large semicircular submarine bar 5 or 6 miles (8 or 10 kilometers) off the Golden Gate. This bar, much of which is barely 35 feet (11 meters) below the surface, consists largely of sand carried northward along this stretch of coast by the same current that built Ocean Beach but which was kept well offshore by the strong ebb tides.

The presence of a submarine canyon in Monterey Bay presents a unique problem. Many such submarine canyons are presumably eroded under water by bottom-hugging currents made heavy by clouds of fine sediment introduced by large rivers from the land. Thus, the Hudson submarine canyon lies off the Hudson River, and the Congo submarine canyon lies off the Congo River. But the Monterey submarine canyon has no major river entering the sea nearby. In contrast, there is no submarine canyon opposite the Golden Gate through which the drainage of the Great Valley has been escaping for much of the last 2 million years. We can understand, therefore, why the suggestion was long ago made that the master drainage, in pre–San Francisco Bay time, flowed south down the San Francisco Bay–Santa Clara lowland to empty into Monterey Bay opposite the head of Monterey Canyon. As a matter of fact, however, we do not have enough evidence at present to know exactly what did take place here. It was thought for a time that there might be a submarine canyon off the Golden Gate, buried under the floods of sediment that were drifted out of the bay or washed along the coast, but geophysical surveys do not support this view. Perhaps a submarine canyon that once formed outside the Golden Gate has moved north along the San Andreas Fault. Or perhaps, in the recent geologic past, the drainage of at least the southern part of the Great Valley filtered through the Southern Coast Ranges and reached Monterey Bay from the south.

THE FUTURE

And what of the future? The future development of the landscape of Middle California will probably not differ significantly from that of the past.

The Sierra Nevada will continue its trapdoorlike rise as a result of numerous small uplifts along the lofty eastern scarp. The westerly flowing streams will continue to deepen their valleys but will find it necessary to fill in the many glacial lakes in their path before downcutting can be resumed in these places. The debris resulting from the valley deepening will continue to be carried down into the Great Valley.

The Great Valley will persist as the repository of sediment from the Sierra and Coast Ranges. Whether deposition will give way to erosion will depend on such events as elevation or depression of the lands, rises or falls in sea level, and climatic changes. For example, if sea level falls again in the future, the master drainage flowing seaward through the Golden Gate may entrench itself deeply, and this deepening may be projected throughout the drainage system in the Great Valley. As a result, the flat valley floor may be eroded into a complex of hills. Later, if sea level rises and conditions conducive to deposition return, the hills may eventually be completely buried and a new flat valley floor formed.

The Coast Ranges will continue to be jostled about, both vertically and horizontally. As before, the jostling will deform the sediments accumulating in the adjacent lowlands. Lateral movements along the San Andreas and similar faults will also deform the adjacent ground. At the present time, for example, about 40 miles (65 kilometers) north of Los Angeles, a large elliptical bulge, the Palmdale bulge, is rising along the San Andreas Fault. It is not clear whether the bulge portends a new displacement along the fault or represents a welcome release of accumulated energy.

As long as blocks of the earth's crust keep shifting about, earthquakes will be common. And the shaking will continue to jostle loose some landslides from the steeper slopes. The shaking will also create new sag ponds by jostling the loose alluvium of valley floors and creating new depressions.

San Francisco Bay, like many of the smaller bays along the coast, would normally be expected to fill up with sediment. But other events may intervene before this happens. Much will depend on whether sea level continues to rise as present-day glaciers melt, or whether it will drop as we turn the corner into a new ice age. If the former proves to be true, the sea will submerge more and more of the bay lowland, requiring the construction of ever higher dikes to keep the sea from flooding the marginal plains. If, on the other hand, sea level begins to subside as more of its water is locked up on land as glacial ice, the sea will withdraw from the bay lowland and the old California River will once again flow out the Golden Gate to a shoreline off the present coast. These changes will be very slow, however, and are matters of concern only to future generations.

Of more immediate concern are the impacts of man's activities. In 1850, the bay covered an area of 680 square miles (1760 square kilometers). It has since shrunk by 40 percent to 420 square miles (1090 square kilometers). About 260 square miles (670 square kilometers) of shallow water and marshes have been "reclaimed" for agriculture and salt ponds and, more recently, for development. If this rate were permitted to continue, in a few generations the bay would have been reduced to a network of tidal channels, the rest having been converted to dry land. Fortunately, the San Francisco Bay Conservation and Development Commission was recently established, and it has the responsibility to insure wise development of the bay shores and to provide generously for natural preserves.

East of the bay is the vast delta of the Sacramento and San Joaquin rivers. Plans are now afoot to pass much of the water of the Sacramento River around the delta through a peripheral canal and divert it to southern California. Loss of water to the delta plain, however, could lead to desiccation of the ground, lessened agricultural suitability, greater pollution concentration from agricultural wastes, possible undesirable impacts on the fishing industry, and invasion of the fresh ground water by salt water. Inasmuch as wells supply much of the fresh water used domestically and for irrigation, the saltwater threat is not to be taken lightly. As for the impact on fishing, the ecology of San Francisco Bay is delicately adapted to a complex regimen in which both fresh water and salt water play a part. During floods, fresh water may mantle much or all of the salt waters of the bay. Actually, in 1860, fresh water extended seaward as far as the Farallon Islands. It is as yet unknown what effects lessened freshwater flow would have on fish such as salmon and striped bass, which live in the ocean but pass through the channels of the freshwater delta to spawning grounds upstream.

Man influences the delta in another way. The tributaries of the Sacramento and San Joaquin rivers spread fanlike over the delta area. The major rivers and their larger tributaries have prominent levees, and these enclose large low-lying areas between them. Formerly, in times of flood, the rivers broke through these levees and flooded the basins, relieving the flood pressures downstream. As long as the basins (called "islands" because they are surrounded by channels on all sides) were uninhabited, they raised no particular problem. But in order to make use of these basins, man has built the levees higher and reinforced them. The basins were then pumped dry, which caused their floors to settle further. Some are now as much as 17 feet (5 meters) below sea level, not unlike parts of The Netherlands. The dikes are maintained at public expense. At present, when the levees are breached and the basins flooded, appreciable monetary losses result from the destruction of crops and the damage to buildings. Whether it is feasible to continue to reclaim these flooded basins is a subject of debate.

In the interior valleys of the Coast Ranges, and in the Great Valley itself, uncontrolled pumping of ground water has caused subsidence over wide areas. In Santa Clara Valley south of San Francisco Bay, subsidence averages 3 feet (1 meter) over an area of 100 square miles (260 square kilometers), with a maximum of 8 feet (2.5 meters) in part of the area. Subsidence on a larger scale has occurred in parts of the Great Valley. During the past 50 years, the ground has subsided 3 to 30 feet (1 to 9 meters) damaging wells, canals, bridges, and other structures, and impairing the fields for use of large equipment.

Man has even added to the landslide hazard by building, excavating, or dumping earth on unstable slopes. Costs run into many millions of dollars annually in California.

So, for the present and immediate future, man seems to have a greater impact on the landscape of Middle California than nature has. In the long run, however, fluctuating climates, rises and falls in sea level, and uplift and foundering of the lands will determine the major changes in the evolution of the landscape of Middle California.

SELECTED GLOSSARY

The Glossary provides a more detailed explanation than the text for selected terms.

Alluvium: Sediment deposited by streams on floodplains and deltas and as fan-shaped forms at the mouths of mountain valleys. May consist of clay, silt, sand, or gravel, or combinations of these.

Ash, volcanic: Small particles hurled aloft during volcanic eruptions. We herein include as ash the somewhat larger particles known as cinders. Ash may be spread outward for great distances, especially to leeward of the source. When consolidated becomes the rock, tuff.

Asthenosphere: See plate tectonics.

Batholith: A large body of molten rock, measured in miles and formed by fusion miles below the Earth's surface. When solidified, may be exposed at the surface by deep erosion, commonly in the cores of mountain ranges. Multiple batholiths may form entire mountain ranges such as the Sierra Nevada.

Chert: A dense, white, black, or colored porcelainlike rock that occurs in isolated nodules or thin layers. The reddish, layered cherts of Middle California were precipitated on the sea floor at great depths.

Cross-bedding: A complicated type of layering in sediments in which sets of dipping layers are truncated and overlain by other sets at different angles. In fluvial cross-bedding, a layer of inclined beds is sandwiched between essentially horizontal layers. In eolian cross-bedding, the inclined sets of beds are truncated and overlain by other inclined beds at different angles and there is no horizontal layering as in fluvial cross-bedding.

Crush zone: A zone of crushed and broken rock generally created by repeated faulting along closely parallel faults.

Dike, artificial: An embankment, generally earthen, built along a watercourse to confine the stream to its channel and/or to protect adjacent areas from flooding.

Dike, igneous: A tabular sheet of igneous rock which forced its way, when molten, into a crack in the neighboring rocks and solidified there below the surface of the Earth. Because of the resistance of many dike rocks, they stand as walls when exposed by erosion.

Earthquake: A sudden trembling of the Earth most commonly caused by faulting, the sudden displacement of segments of the Earth's crust.

Erosion: The loosening, dissolving, or wearing away of exposed rocks or other earth materials by running water, glaciers, wind, waves, or currents.

Estuary: The tidal mouth of a river valley where fresh and salt water intermingle.

Exfoliation: The process by which thin sheets of rock, often shell-like, are loosened from bedrock exposures. Much exfoliation is due to weathering processes that cause expansion and liberate sheets from less than an inch to several inches thick. Exfoliation on a more massive scale results when deep erosion relieves the downward pressure on once deeply buried rocks and results in large topographic domes as in the Sierra Nevada.

Fault: A fracture in rock along which there has been displacement of the two sides. Displacement, at any one time, may range from a fraction of an inch to tens of feet. Recurrent movements may lead to accumulated displacements of many miles.

Fumarole: A vent, generally in volcanic areas, from which gases and vapors are emitted.

Geologic calender: A record of the divisions of geologic time including eras, measured in scores or hundreds of millions of years, and smaller subdivisions known as periods and epochs. A geologic calender is presented in the text.

Geology: The study of the Earth, its origin, materials, the processes that modify it, and its history through time. A *geologist* is a specialist in this field.

Geothermal: Pertains to the Earth's internal heat. The increase of temperature with depth, within the range of man's observations, averages about 2.5 degrees Celsius (2.5°C) per 100 meters or less than 1 degree Fahrenheit (1°F) per 100 feet. Heating of ground water by geothermal heat gives rise to hot springs and geysers, or—if only vapors are produced—to fumaroles.

Igneous rock: A rock that was originally molten. Molten rock in depth is known as *magma.* Magma may rise into higher levels of the crust and solidify without reaching the surface. Such igneous bodies are referred to as *intrusive.* Magma may also reach the surface and pour out as lava, an *extrusive* form. *Granite* is the most common intrusive rock, whereas *basalt* is the most common extrusive type. *Rhyolite* is a light-colored lava. The upper, more rapidly chilled part of a rhyolite flow is often a dark glassy rock known as *obsidian.* The very top of the flow, the lava froth, solidifies as the highly porous gray rock, *pumice.*

Levee: Natural embankments along the margins of stream channels on floodplains. Constructed of overbank flood deposits.

Magma: See Igneous rock.

Marine sediment: Sediment deposited in the sea, as contrasted with terrestrial sediment deposited on land.

Metamorphic rock: Rocks derived from pre-existing rocks in response to marked changes in temperature, pressure, or chemical environment. The new rock may differ from the original in mineral assemblage, chemical composition, and internal structure. Examples cited in the text are slate, phyllite, schist, marble, quartzite, gneiss, and serpentine rock.

Mudflow: A slurry of saturated clay, silt, or both, which moves rapidly downslope often spreading out over adjacent lowlands. Many alluvial fans contain layers of mudflow material. Mudflows are common on slopes of volcanoes where heavy rainfalls saturate the abundant ash on the steep slopes.

Orogeny: Mountain-making, involving folding and faulting, and commonly metamorphism of the affected rocks.

Pediment: A sloping erosion surface stretching outward from the foot of mountain or plateau escarpments in semiarid or arid environments. The surface is commonly veneered with a mantle of alluvium.

Placer deposit: A mineral deposit formed on the surface by mechanical concentration of mineral particles. Concentration by streams is most common, but wind, waves, and currents are also important.

Plate tectonics: Global tectonics (crustal deformation) involving huge plates of the Earth's crust moving about over the *asthenosphere,* a viscous, partially molten layer below the Earth's crust. The plates grind against each other like ice floes and account for much of the Earth's orogenic activity.

Pyroclastic materials: Materials exploded into the air during volcanic eruptions. Ash represents the finest particles. For convenience, we use the term ash to include the next larger particles, cinders. Ash and cinders solidify to form the rock, tuff.

Radioactivity: The spontaneous decay of the atoms of certain elements. During decay, certain component particles of the atom are lost and heat is generated. The rate of decay for a particular radioactive element is constant thereby providing a basis for determining how long ago the element was formed. Thus, in about 5700 years, half of the radioactive carbon atoms in organic materials reverts to non-radioactive nitrogen. In the next 5700 years half of the remaining radioactive carbon reverts to nitrogen, and so on indefinitely.

Rock: Any naturally occurring, hard, consolidated material composed of one or more mineral species welded together as in igneous and many metamorphic rocks, or consisting of grains or fragments derived from pre-existing rocks and consolidated by natural processes.

Sag Pond: A small body of water occupying a depression created by irregular settling of the ground during earthquakes.

Sediment: Solid fragmental material, inorganic or organic, that originates from weathering or erosion of rocks and which is transported by water, air, or ice and eventually deposited.

Substances transported in solution are also precipitated as sediment. Common fragmental sediments are clay, silt, sand, and gravel, whereas many limey muds are products of chemical precipitation.

Sedimentary rock: Rock formed by the consolidation of sediments. The rock equivalents of clay, sand, and gravel are shale, sandstone, and conglomerate respectively. Limey muds solidify to limestone.

Subduction: The descent of one crustal plate under another in plate tectonics. The edge of the descending plate eventually reaches depths at which fusion and the generation of magma may take place.

Tectonics: A general term referring to deformation of the Earth's crust by faulting, folding, and/or metamorphism, and commonly resulting in mountains.

Terrestrial sediments: Sediments deposited on land as opposed to marine sediments laid down in the sea.

Topographic relief: The difference in elevation between the high and low spots in a particular landscape. In deeply eroded mountains and plateaus the relief may be thousands of feet; in plains areas, it may be only a few feet.

Transform fault: A fault cutting across oceanic ridges and displacing them laterally.

Trellis drainage: A pattern in which major tributaries on opposite sides of a main stream are aligned with each other and join the main stream at right angles. Their tributaries in turn enter at right angles. The overall pattern resembles a trellis latticework such as gardeners use to train vines.

Tuff, volcanic: A rock consisting of solidified volcanic ash and cinders. *See also* Volcanic ash.

Weathering: The physical disintegration or chemical decomposition of rocks in contact with the atmosphere. The expansion of ice in cracks, temperature changes, the pressure of growing roots, and other processes result in physical disruption of rocks. Rainwater, containing elements picked up on passage through the air, assists in chemical decomposition of susceptible rocks. Rainwater with dissolved carbon dioxide, for example, is a mild acid.

SELECTED REFERENCES

Bailey, E. H., editor, 1966, Geology of Northern California: California Division of Mines and Geology, Bulletin 190, 28 articles, 7 field trip guides, 508 pages.

California Division of Mines and Geology, 1958–1969, Geologic Atlas of California, 27 sheets, 1: 250,000, prepared under the direction of C. W. Jennings.

Dickinson, W. R., editor, 1974, Geologic Interpretations from Global Tectonics with Applications for California Geology and Petroleum Exploration: San Joaquin Geological Society, Bakersfield, California, 9 articles, 70 pages.

Donnelly, J. M., F. E. Goff, B. C. Hearn, Jr., and R. J. McLaughlin, 1977, Field Trip Guide to the Geysers–Clear Lake Area: Cordilleran Section, Geological Society of America, 2 articles, 56 pages.

Greene, H. G., 1970, Geology of Southern Monterey Bay and Its Relationship to the Ground Water Basin and Salt Water Intrusion: Open File Report, U.S. Geological Survey, Prepared in cooperation with the California State Department of Water Resources, 50 pages, maps.

Howard, Arthur, 1962, Evolution of the Landscape of the San Francisco Bay Region: University of California Press, 72 pages.

Jenkins, O. P., director, 1948, The Mother Lode Country: California Division of Mines and Geology, Bulletin 141, 10 articles, 164 pages.

Jenkins, O. P., director, 1951, Geologic Guidebook of the San Francisco Bay Counties: California Division of Mines and Geology, Bulletin 154, 32 articles, 392 pages.

Nilsen, T. H., editor, 1977, Late Mesozoic and Cenozoic Sedimentation and Tectonics in California: San Joaquin Geological Society, Bakersfield, California, 14 articles, 145 pages.

Wahrhaftig, C., and J. H. Birman, 1965, The Quaternary of the Pacific Mountain System in California: in The Quaternary of the United States, H. E. Wright, Jr. and D. C. Frey, editors, Princeton University Press, pp. 299–340.

INDEX

Numbers in *italics* indicate figure pages. Photos referred to separately.

Compositor:	Viking Typographics
Printer:	Consolidated Printers
Binder:	Consolidated Printers
Text:	Times Roman
Display:	Lydian Medium
Paper:	50 lb. P&S Offset